SURVIVING GLOBAL WARMING

SURVIVING GLOBAL WARMING

WHY ELIMINATING GREENHOUSE GASES ISN'T ENOUGH

ROGER A. SEDJO

 Prometheus Books

59 John Glenn Drive
Amherst, New York 14228

Published 2019 by Prometheus Books

Cover design by Jacqueline Nasso Cooke
Cover image © D. Hurst / Alamy Stock Photo
Cover design © Prometheus Books

Inquiries should be addressed to
Prometheus Books
59 John Glenn Drive
Amherst, New York 14228
VOICE: 716–691–0133 • FAX: 716–691–0137
WWW.PROMETHEUSBOOKS.COM

23 22 21 20 19   5 4 3 2 1

Library of Congress Cataloging-in-Publication Data

Names: Sedjo, Roger A., author.
Title: Surviving global warming : why eliminating greenhouse gases isn't enough / by Roger A. Sedjo.
Description: Amherst, New York : Prometheus Books, 2019. | Includes index.
Identifiers: LCCN 2018056718 (print) | LCCN 2018060088 (ebook) | ISBN 9781633885295 (ebook) | ISBN 9781633885288 (hardcover)
Subjects: LCSH: Greenhouse gas mitigation. | Greenhouse gases. | Climate change mitigation.
Classification: LCC TD885.5.G73 (ebook) | LCC TD885.5.G73 S43 2019 (print) | DDC 363.738/74--dc23
LC record available at https://lccn.loc.gov/2018056718

Printed in the United States of America

I want to dedicate this book to my wife, Ruthy,
who stayed steady during some hard times. Thanks sweetheart.

CONTENTS

ACKNOWLEDGMENTS

There are many people who are responsible for this book. Included are those at Resources for the Future (RFF), various US government agencies, and international groups, including especially my colleagues at the UN Intergovernmental Panel on Climate Change who worked with me on the various Assessments. As I have been working on this book directly, my wife, Ruthy, provided assistance and tolerated the various difficulties and stresses associated with this effort.

Upon my retirement from RFF, I had little initial interest in taking on another book. However, in retrospect such an endeavor was not unlikely as the climate issue continued to fester and grow as a social issue and I did have both a long history in the topic and a number of unresolved related questions in my own mind. Friends and acquaintances continued to raise the issue. Professor Dean Lueck, now at the University of Indiana but formerly at the University of Arizona, continued to raise technical questions and invited me to give presentations, including a number at his Natural Resources classes.

Other individuals I wish to thank include Franz Fauley and Peter Rodman, both recent friends, who stimulated my renewed interest with their questions as interested lay persons. My former colleague A. Clark Wiseman was most helpful, both with focused questions and with assistance in articulating more precisely some of the climate warming issues. Also, I thank my agent, Claire Gerus, who successfully operated through the process of finding the right publisher, and then "pushed" me into new climate territories as I moved through the completion of my writings. Finally, I wish to thank Prometheus

Books and their staff and especially Steven L. Mitchell, editor, who provided guidance and editorial oversight throughout the process. Nevertheless, any errors that remain are solely those of the author. The book was not sponsored by RFF, and I am wholly responsible for its substances.

INTRODUCTION: CLIMATE CHANGE

WHERE ARE WE NOW?

R ecently, the *New York Times Magazine* published an article, "Losing Earth,"[1] suggesting that the whole global change crisis could have been avoided if we simply had signed an international treaty in the 1980s outlawing the buildup of human-generated greenhouse gases (GHGs). As one who worked in this area during the 1980s, I find the naiveté of this view palpable. As noted below, even some of the fathers of the GHG warming hypothesis (e.g., Roger Revelle, an early researcher of human-caused warming) questioned the extent of its operational efficacy early on. Uncertainty existed within both the research community and in the government agencies, to say nothing of the skepticism within the US Senate, which would need to ratify any such treaty.

As this book demonstrates, warming is not new to the earth. Observe Bruce Gardner's comments in the early 1990s, where he notes that time would determine whether increased human GHGs could have any significant impact on Earth or was simply a passing occurrence with negligible effect. Government research shops in agencies such as the EPA were still focusing on the acid rain problem and ozone issues. Many of these researchers were beginning to be redirected to climate change projects in the late 1980s and early 1990s.

The notion of the *New York Times Magazine* article seems to be that if the US signed a treaty, the problem was solved. This reminds me of the mythical King Canute, who while standing on the seashore, ordered the tide to stop. The only outcome was the king got his feet wet. Even in the 1980s, the US was just one among most of the world's countries emitting GHGs. Many of those countries, including China, South Korea, Taiwan, and Indonesia, were almost surely unwilling to give up their focus on economic growth and to replace fossil fuel power with the then underdeveloped and primitive renewable power of the time.

In the 1980s, the serious alternatives to fossil fuels were few and far between. Operational wind and solar farms were still decades away. Photovoltaics could produce energy from the sun, but only at several times the cost of fossil fuels. Indeed, even in the late 1990s, an important article[2] argued that society should spend another couple of decades studying the warming phenomenon and developing responses before we, like Don Quixote, blithely charged the little understood climate change windmill.

In the fall of 2017, within three weeks, three major hurricanes—Harvey, Irma, and Maria—brought havoc to the Gulf and East Coasts of the United States, as well as to US protectorate, Puerto Rico. Although some might challenge the storms' direct relationship to climate warming, it is indisputable that the warmer the air, the more water it can hold (4 percent per degree Fahrenheit). Combined with the warming waters of the Gulf, the hurricane-force winds resulted in unprecedented volumes of water dumped on the earth from Hurricane Harvey.

Similarly, Hurricane Irma brought near-record winds to Florida's East Coast. Ultimately, however, the worst hit was Puerto Rico, which sustained long-term destruction that could take years to overcome.

It's clear that our planet is under siege, and despite the resistance of a small percentage of naysayers, alarms from both sides of the political aisle are being voiced over the apparent intensification of climate change. (Indeed, articles are beginning to appear in such media as *The New Yorker*, questioning how habitable Planet Earth will be in the year 2100[3]—a telling departure from the magazine's usual focus on social/political fare.)

Adding to this uneasy mix is the fact that government scientists have prepared a climate report that turns out to be at odds with the position of the Trump administration. Among other things, both camps differ on the cause and extent of global warming, as well as its potential long-term effects. Nevertheless, climate change has become one of the most difficult issues facing humanity. Former Vice President Al Gore has called climate change an "inconvenient truth." It's now becoming far more than "inconvenient," however, as many feel our species is at a tipping point.

Al Gore has directed attention to GHG emissions as the cause of global warming and builds on the worldwide environmental disasters that scientists project. Although I share his concerns, I tend to disagree that this is the whole story. True, Gore's is the dominant view, which is captured in the recent US Global Change Research Program's Climate Science Special Report. However, I believe this view is incomplete. The Global Change Report[4] states, "It is extremely likely that more than half of the global mean temperature increase since 1951 was caused by human influence on climate." Thus, although GHGs explain much of current warming, the report concedes that it does not explain a large portion of the change. In reality, climate history reveals a global history of variable climate. Well-known is the planet's experience with recurring ice ages. Less well-known is the occurrence of a number of earlier warmings since the end of this last ice age. These are recent enough to have left their mark in human history but still early enough to predate human involvement and so cannot be attributed to human activities. Some of these warmings do not seem to be related to carbon or GHG emissions at all, but still must be factored into the entire scientific exploration to help us understand what contributes to climate change.

One could describe exploring climate change as similar to peeling the proverbial multilayered onion. Below each layer lies another question. Is climate change real? Is the temperature of the earth actually rising? If so, are humans in some manner responsible? If so, are GHGs the driver? Questions also arise as to how best to address the concerns. If GHGs are the problem, can humans control

GHG emissions adequately to stabilize global temperatures? When questions are proposed this way, we begin to discover a great deal of unevenness in what most experts believe about the above questions.

Nevertheless, as the world community chooses to follow Gore and the GHG theory of causation, it has responded by marshaling its collective resources, largely through UN leadership, to stop a warming disaster by preventing the emission of GHGs. Alas, this approach fails to factor in that human GHG emissions are only the human dimension in what appears to be a broader global climate change issue. In fact, solar events, tectonics, volcanic activity, and ocean currents can dramatically affect climate and likely have in this planet's past.

Because there is a lack of attention to natural forces affecting climate change, the popular belief is that simply stepping on the GHG emission brakes will lessen, or "mitigate," the buildup of GHGs. Thus, there is a movement to minimize emissions by minimizing the use of fossil-fuel energy.

Although I support the mitigation effort, I strongly believe that such an approach by itself is inadequate to the task before us. Marshaling world governments to reduce the impact of climate change has focused on restructuring the energy industry away from fossil fuels and toward renewables. However, this solution is still problematic. If the warnings are overblown, as some believe, the damages from fossil fuels would be modest, and the benefits of fossil fuel reduction would be minimal. If the warming is also driven by natural forces, some of the resultant negative effects would continue even without the impact of added human-driven GHGs.

After years of studying the evidence, I am convinced that the nations on this planet are taking the wrong approach with their almost exclusive focus on preventing GHGs to address climate change. It is now recognized that some warming has become inevitable, and despite the huge expenditure of resources on controlling fossil fuels and GHG emissions, any plausible amount of prevention will still be inadequate to reduce greenhouse gases enough to stabilize temperatures within the desired range.

Given the present reality of GHG emissions, we must take a practical and more comprehensive approach and address *both* their existence, and the climatic damages that will undeniably occur. If unusually high temperatures or other phenomena do indeed occur, people will need to employ major efforts toward adapting to and managing the resulting effects.

In this book, I propose that we include adaptation as a practical addition to the current approach of mitigation. I will discuss major areas like the coastal zone, the agricultural sector, and ecological resources as likely to experience extreme damage. I will explore some of the approaches that can be used for adaptation and damage control management in each area.

I believe it is important to become aware of this planet's climate history and how we human beings can be affected by climate warming generated by natural variation. Natural warming is not unique to our time; in fact, the earth's climate system has been inherently unstable. From the hundreds of millions of years since the first ice age to the current interglacial warming period beginning about eleven thousand years ago, climate change has been part of the earth's climate legacy. This includes the period of Viking colonization of Greenland, one thousand years ago, and the subsequent Little Ice Age that only ended as recently as the 1800s.

If natural forces contributed regularly and significantly to the earth's warming as recently as two centuries ago, how can we assert that natural forces are not major contributors today? Climate change, including warming, will continue to be driven by natural forces, with concomitant disruptions of human habitats. Therefore, we humans must use a combined approach, so we can address and adapt to warming from both natural and man-made factors and are prepared for whatever damage management is required.

Humankind is at a crossroads. Is there more than one source of global warming that needs to be addressed? How can a variety of political systems in the United States and worldwide deal with all the causes and results of this phenomenon?

For almost four decades, I have been involved in research on resources, the environment, and climate as a member of Resources for the Future, an independent, nonpartisan resource and environmental think tank in Washington, DC. Simultaneously I worked with the Intergovernmental Panel on Climate Change (IPCC), and along with some of my IPCC colleagues, I shared the Nobel Peace Prize in 2007 for my participation in writing three volumes of the UN Assessment on Climate Change (1995, 2001, 2006). Al Gore received the Peace Prize that year for his individual work.

At present, the UN's attitude toward climate change focuses myopically on the conclusion that human-generated GHGs are solely at fault. I call it "Plan A," or "The Mitigation Solution," since it focuses on minimizing or eliminating GHGs to lessen the impact of global warming. The present focus of the global community on mitigation implies the need for an international bureaucracy and large-scale, centralized cooperative action. However, this raises questions about how individual countries can progress if all are not in sync with an adequate budget and a uniform approach. "Plan B: The Adaptation Solution," as proposed in this book, enables affected countries and local areas to apply specific, individual preparations and responses to unique national and local climate challenges.

Ultimately, I propose that we must implement "Plan B: Adaptation" as a backup, together with "Plan A: Mitigation." Only then can we gain confidence that humanity will, indeed, be up to the challenge of successfully coexisting with the reality of climate change.

CHAPTER 1

AL GORE AND THE GREENHOUSE GAS THEORY: PLAN A

Late in the 1980s, both the US government and the research community began to pay more serious attention to issues related to climate change, and particularly global warming. It was well-known that the earth's atmosphere included a number of so-called "greenhouse gases" (GHGs). They are the gases in the atmosphere that absorb and reflect radiation, trapping the sun's energy as heat and thereby maintaining the warmth of the earth. The concept that changes in the composition of the atmosphere could change climate go back at least as far as Tyndall in the 1860s, and Arrhenius's seminal paper in 1896.[1] In addition to water vapor, the gases included carbon dioxide, methane, and nitrous oxide, as well as some other gases less important for heat-absorbing purposes.

In 1989, the US Department of Agriculture held a meeting featuring a number of invited papers on the anticipated impact of climate change on agriculture. Presiding over this meeting was the late Dr. Bruce Gardner, a senior Department of Agriculture official, who informed us researchers that the coming years would see us deeply immersed in climate research. If climate change was not "real," he noted, it would probably take many years to confirm its absence. If it were real, climate research would definitely gear up for many decades ahead.[2]

Gardner was right, and since warming appears to be continuing and doubts of its authenticity diminish, research into the nature of this phenomenon and its implications has greatly accelerated. The objective: to better understand what drives global change as well as what the implications of climate change are likely to be for the future of Earth's inhabitants. Once understood, the next step is to identify actions and policies that can prevent warming and the damage associated with climate change—no small task, indeed!

To date, one view and one approach has gained prominence—and to a great extent, this is due to its most prominent supporter, former Vice President Al Gore, who has worked tirelessly to advocate and convince those in power, as well as ordinary citizens, to abandon fossil fuels in favor of alternate energies (e.g., wind, solar, etc.).

GORE'S VIEW

The bare bones of today's primary climate change view are described in Gore's bestseller, *An Inconvenient Truth*.[3] As described in both the book and later the film, Earth's temperature is rising, due mostly to carbon dioxide and other greenhouse gas emissions released through the burning of fossil fuels, as well as deforestation for urban development, which releases certain GHGs from trees and soils. The implication is that we humans can save ourselves, and our planet, by being less greedy in our use of energy and, with the judicious use of technology to reduce energy use and to find alternatives to fossil fuels, in the form of renewable energy.

Gore attributed much of his interest in climate change to his Harvard teacher Roger Revelle, who, back in the 1950s, had been an early researcher of the link between carbon dioxide and global warming. However, Revelle became more skeptical of the power of this relationship as he grew older. In a conversation I had with Revelle in the late 1980s at Resources for the Future (where he was visiting for a meeting of RFF's Climate Resources Program Advisory Committee), he confided that he was concerned that Al Gore was possibly overstating the confidence we could have in the Carbon/ Greenhouse Gas view. In time, this dispute between teacher and pupil made its way into print in an article Revelle authored with others in the magazine *Cosmos* (1991), stating: "The scientific base for a greenhouse warming is too uncertain to justify drastic action at this time. There is little risk in delaying policy responses."[4]

While Revelle's concerns launched similar criticisms from other scientists, the public continued to support Gore's view. Subsequently, rivals of the GHG embracers were tagged with the term "deniers" and cast as laggards amid the growing urgency for the world to "do something!" to save our planet.

Coming out of such debates, theories began to arise regarding the underlying causes of not only warming, but also of the growing environmental movement. Robert Nelson of the University of Maryland, for example, argues that environmentalism has a strong religious flavor. As the pressures of life cause us to feel under siege, he contends that we view ourselves in need of secular redemption.[5]

Who or what will save us? Many believe we should cast off destructive environmental activities, including the use of fossil fuels, and look to technology or a simpler life for salvation. This would, of course, be a massive undertaking, given that the causes of warming are still under investigation and eliminating fossil fuels would doubt-less generate substantial lifestyle changes. Nevertheless, environmentalism has an increasingly powerful influence on many of us, and climate change simply adds to the list of our concerns.

Despite these criticisms and others, *An Inconvenient Truth* clearly describes the GHG warming hypothesis, and has achieved almost universal acceptance worldwide, particularly among the leaders of major nations. The almost consensus acceptance of the Paris Climate Agreement attests to the widespread acceptance of the GHG view of climate change.[6] Most surveys place acceptance of human-driven climate change at nearly 90 percent.[7] Surprisingly, the acceptance of Gore's concept of climate change and warming is somewhat lower in the United States than elsewhere. In fact, at least one-third of the

US population questions the basic GHG climate warming notion.[8] In addition, some members of the American scientific community are skeptical of this view and have pushed back, regarding this view as overly simplistic and as ignoring a number of relevant facts.

A theory of climate change must, some believe, provide an understanding not only of humankind's impact on climate, but that of natural events, such as historical climate change and the global climate cycle driven by nonhuman forces. Most contemporary research, they note, has focused on human-generated agents, ignoring much of the earlier climate change prior to the advent of significant human activity.[9]

Let me be clear: Gore's view that GHGs have caused warming certainly has merit. It follows quite closely the contemporary observed climate behavior. Scientists have known since at least the early part of the twentieth century that the gaseous composition of the earth's atmosphere can affect the amount of heat from the sun's solar rays captured on the earth. If there is a significant increase in the fraction of the earth's atmosphere consisting of carbon dioxide, methane, and other heat-capturing greenhouse gases, the solar energy captured should definitely increase the earth's temperature.[10] The carbon-GHG perspective gives science a hypothesis for the capture of increased solar energy and an explanation for warming. Empirically, scientists find the amount of carbon dioxide and other GHGs in the atmosphere *is* increasing. In the pre-industrial period, the portion of carbon dioxide in the atmosphere was estimated at 280 parts/million. Today, it is about 410 p/m.[11]

The portion of the atmosphere that is made up of GHGs has been watched very closely since 1958, when monitoring pristine atmosphere gases from a previously unspoiled location on Mauna Loa in Hawaii showed a progressive increase in the carbon dioxide fraction of the atmospheric gases.[12] Science now has actual empirical observations that match temperatures with GHG levels. These observations show a temperature warming that is loosely correlated with GHGs and generally consistent with the GHG hypothesis.

Hence, the contemporary case for the human carbon-GHG climate driver is, without a doubt, believable.

PUSHBACK

The pushback on the GHG view comes from a small number of scientists and a larger number of skeptical laypersons. At one level, there are scientists who view the correlation between GHGs and temperature as too simplistic. After all, just because the two are correlated does not establish that the former caused the latter. Many other factors could be correlated with climate change, and may simply be correlated with time. Additionally, measuring the earth's temperature over time to see if it is actually rising is no easy matter. Although ice cores can be used to measure carbon levels at specific locations from earlier periods, measuring temperatures is more difficult. Contemporary measurements must be taken in some representative way across the entire globe if we are to reach what might be termed the "average" global temperature.

It is well-known that even when temperatures are rising in most places, they are not rising everywhere. Hence, it is hard to know when a truly representative temperature is found. Urban heat belts, warmer areas caused by features within urban centers—streets, cement walks, and other features that absorb more heat than natural settings—are known to include local distortions and so do not reflect representative global warming. Therefore, some measures must be removed or adjusted from the analysis. Temperatures over the oceans must accurately and consistently be included, and over the past few decades, there have been changes in how these are measured.[13]

An alternative approach is to measure the temperature above the earth using satellites. However, even this approach has its own set of complications and inconsistencies. Historic temperatures are even more difficult to obtain with no consistent recorded global temperature estimates until, at the earliest, beginning after the mid to late

1800s. Very early temperature measurements, preceding temperature gauges, are often arrived at by studying changes in crop types, tree elevations, tree rings, and locations of plant types and seeds. Certain chemical tracers in shells also link closely to temperature at the time they were created. For example, by studying oxygen isotopes, scientists can deduce temperature conditions at the time that the shell was formed.[14] Nevertheless, as remarkable as these techniques are for estimating the general level and direction of temperatures in a long gone-by period, reliable, precise historical temperature data amenable to statistical testing is almost always nonexistent or at best crudely constructed.

Much of the case for GHG climate change, and particularly some of the draconian future scenarios, are based on projection models and the correlation between the short-term buildup of GHGs and climate warming. For example, the links between the levels of GHGs, future temperatures, and sea-level rise are highly uncertain. How rapidly will land-based glaciers melt, and will future snows offset much of that melting?

The short-term correlation between GHGs and temperature is rather weak. The notion of a tight correlation between GHGs and consistently rising temperatures is being challenged due to regular departures from the trends. For example, in the period 1998 to 2015, when GHGs were very closely monitored, climate change skeptics maintain that temperatures did *not* rise higher than the 1998 level for nearly two decades, despite the continuing and significant rise of atmospheric carbon dioxide.[15] However, temperatures did begin to rise after this period. At the very least this pattern suggests that yearly global temperatures are *not* tightly tied to an annual GHG driver. In fact, some now argue that the correlation, at best, indicates a relationship between the *longer-term* trends, perhaps multidecadal, in GHGs and rising temperatures.

Scientists have been seriously addressing the forces that might be generating climate change for only a few decades. Since this is a relatively short time, the notion that climate behavior is in any way "settled" is, to me, premature. Do we seriously believe that we know so much about climate change today that in fifty years climate science will be the same? Even something as well established as Newtonian physics, which seemed "settled" for centuries, has proved inadequate for certain problems. In addressing bodies approaching the speed of light, Einstein's Theory of Relativity has supplanted it. Science, and certainly climate science, must be viewed as provisional at best and continually subject to revision.

Finally, there is the IPCC (Intergovernmental Panel on Climate Change) itself. In maintaining the importance of a human factor in the warming system, the IPCC stated that "more than half of the observed increase in global average surface temperature from 1951 to 2010 was caused by the anthropogenic increase in greenhouse house gas concentrations . . ."[16] Fine, but what is the source of the remaining other half of the warming? The other half is what the IPCC and science has yet to explain, leaving a more comprehensive explanation of climate change yet to be discovered.

NATURAL FORCES AND CLIMATE CHANGE

Although the focus of climate control is on human-generated GHGs emitted predominantly from fossil fuels, the evidence suggesting an important natural component to warming is growing. We know that once ice covered much of Earth. Eventually, the early ice cover declined, but it was subsequently replaced by a series of perhaps five glacial periods. The extent of the glacial ice within the current ice age tends to vary, typically lasting around one hundred thousand years, with occasional interruptions of tens of thousands of years called interglacial periods. At the beginning of our current interglacial period, so much water was tied up as ice that sea levels have risen some 125 meters in the past twelve thousand years. The drivers of the comings and goings of large-scale changes in the earth's climate, which are manifested as ice ages and interglacial cycles, appear to be fluctuations in the variations in intensity and timing of energy from the sun. These major variations appear to be driven by orbital changes and the influence from the earth's wobbly spin on its axis, which effect the earth's tilt toward the sun. Changes in orbit can affect the position of the earth on its elliptical path, including the angle of the earth's axis and distance of the earth from the sun. These conditions interact with the earth's atmosphere and water to affect the temperature. Such movements could impact both the seasons and amount of solar energy capture.[17] More modest variations in climate within an interglacial period seem to be due to factors such as volcanoes, ocean circulation, solar factors including sunspots, and clouding, as well as the intensity and timing of heat from the sun, which depend also on orbital factors like the earth's inclination to the sun[18] and even the position of the continents. Additionally, of course, the types and prevalence of greenhouse gases, such as carbon dioxide in the atmosphere, clearly affect the earth's temperature and climate.

Scientists have identified about eight periods of distinctive climate change since the beginning of our latest interglacial period. Only five have been in the historical period. Two occurred during in the Bronze Age, perhaps about 1750 BCE and 1250 BCE. The last of these may have precipitated the end of the Bronze Age by creating the mass movements of peoples that led to widespread disruptions in the eastern Mediterranean, the influx of new peoples including the Philistines of the Bible, and the creation of the political states including that of the Jews. Another was the Roman Warming around 200 BCE to 250 CE, covering much of the time period of Pax Romana. Next there was the Medieval or Viking Warming from 950 CE to roughly 1350 CE. Finally, our own contemporary warming, beginning as early as the late 1800s, or perhaps as late as the late 1900s, and continuing to the current period. Although it appears likely that the current warming is driven mostly by human forces, given the earth's experience with climate instability even during interglacial periods, the likelihood that natural forces are also at work appears large. Although the IPCC tends to stress the human factor, the inability of science to explain much more than half of the warming suggests that science still has much more to learn.[19]

RESPONDING TO CLIMATE CHANGE

If carbon dioxide and GHGs are the cause of warming, then fossil fuels are the enemy. Throughout the late twentieth century, fossil fuels were the world's dominant source of energy. They constituted over 80 percent of the earth's energy, and that figure remains essentially the same today, although the mix has shifted away from coal. So, despite the fact that there have been significant efforts to wean societies away from fossil fuels, overall their share of the energy mix has remained stubbornly stable for much of a half of a century. This performance suggests that the task of reducing or at least stabilizing GHG emissions is not easy. It raises the question as to whether our current efforts at reducing such emissions can, in fact, achieve the desired ends.

If warming were solely driven by one factor, such as human releasing of GHGs, the human response would be in principle fairly straightforward. As has been underway for some time, the focus would probably be on controlling the source of the problem. However, a second contributing climate driver complicates matters. A solely GHG mitigation approach would probably be inadequate since it would not address the natural release of such gases into the atmosphere. In conjunction with trying to prevent the event, the release of GHGs, the adaptation approach would involve the anticipation of the event, its emissions and warming, and then an attempt to manage both the event and any associated damages, for example by building seawalls. Of course, adaptation is much less satisfactory than mitigation. But in many cases, successful mitigation might be unattainable. To the extent that warming is driven by forces other than GHGs, as suggested by the historical planetary warmings that apparently occurred without GHGs, mitigation could have only minimal corrective effects. In other words, attempts to reduce GHGs may not be able to stabilize or reduce the rise in global temperatures if the cause is driven by multiple factors. As will be discussed later in this volume, attempts at mitigation may be insufficient and therefore unsuccessful. Finally, there is growing concern that the effects of warming have long lag periods and even stabilizing emissions today will not alleviate long-term global system disruptions. A recent study from Germany's Potsdam Institute suggests that today's mitigation failures will influence the rise in sea levels for the next three hundred years.[20] These climate drivers cannot be undone easily. Adaptation is designed to address damage and, in any event, ought to have a positive effect in addressing warming issues whatever their causes.

CONCLUSIONS

This book discusses measures that might be undertaken to address climate and temperature change. The feasibility and potential of the various approaches is examined. The massive reduction of GHG emissions from fossil fuels would necessitate the identification and use of alternative energy sources as fossil-fuel energy replacements. The substitution of other current energy sources, including nuclear, hydropower, and biofuel sources, is probably limited. Given these limits, the candidates as potential replacements seem to be predominantly renewables such as wind, solar, and perhaps biogenic. These energy sources have their own set of problems, such as intermittency of availability. Furthermore, the existing high levels of GHGs in the atmosphere are likely to have long-term impacts regardless of when stabilization occurs.

For the purposes of the analysis of this book, I accept Al Gore's basic view that carbon and GHGs are important driving forces of today's global warming and the result of human activities. However, I also believe that this viewpoint is incomplete and that generating a remedial solution solely based on significantly reducing or eliminating GHG emissions, as is being attempted at the United Nations through the Paris Agreement of 2015 (Plan A), is inadequate. This book undertakes to explain the problems and limitations with implementing Plan A. Although partially effective, I predict that a Plan A approach will be insufficient to fully halt climate warming. This leads me to suggest that both GHGs and natural variability need to be adequately incorporated into our assessment and our solution. The approach I suggest in this volume is the addition of adaptive techniques that anticipate the climate-generated events and move to minimize the damage and, where practicable, to neutralize GHG effects on atmospheric warming. This approach I call "Plan B."

CHAPTER 2

NATURAL CLIMATE CHANGE: GHGs ARE NOT THE WHOLE ANSWER

THE LONG VIEW

Once upon a time, about two billion years ago, an obscure planet in our own dim solar system began to form a huge glacial cover. The ice covered much, but not all, of Earth for over one billion years. Eventually, the ice cover declined, but it was subsequently replaced by a series of perhaps eight glacial periods.

After periods of relative decline, our current glacial cycle began about 2.7 million years ago, at the start of the Pleistocene epoch. This created an ice cover that still exists today as remnant ice sheets in parts of Greenland, the Arctic, and the Antarctic. In the intervening periods between intensive ice cover, however, phases of "hothouse" warming seem to have occurred.[1] Within this glacial epoch, the ice tended to expand and contract, with major expansions typically lasting around one hundred thousand years, before relatively short interruptions called interglacial periods, which lasted tens of thousands of years.[2] Interglacial periods are typically followed by a renewal of the expansion of the ice. Today we are living in an interglacial period. How long this interglacial period will last is unknown.[3]

Figure 2.1 shows fluctuations of carbon dioxide over both a period of four hundred thousand years, and separately during the last thousand years.[4] Looking at the longer four hundred thousand year period, note that the last sharp upturn of the graph is a projection, not an actual recorded data point, of forecasted carbon dioxide/ GHG levels that has not yet been realized. Note that the time scale for carbon levels is essentially tens of thousands of years, with trough to peak time periods approaching one hundred thousand years. So information for the last thousand years cannot usefully be drawn from this graph due to its scale, and useful comparisons cannot be made for much shorter time periods. The point of this graph is that if the earth's carbon level increases as is sometimes predicted, the levels will reach heights unprecedented in the last four hundred thousand years, with all the implications for warming.

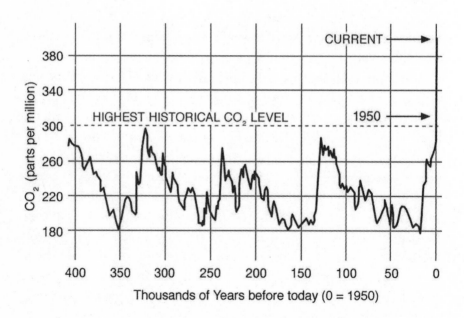

Figure 2.1. (Chart by NASA, data from NOAA.)

A fascinating feature of climate change and warming is the variety of forces that seem to be involved in the changes. The current warming appears to be driven by carbon dioxide and GHGs. However, as I demonstrate in this chapter, a host of earlier warmings appear to have occurred without any serious increase in carbon dioxide or GHGs. I examine some of that history, noting both the differences and the similarities. I concur with the conventional wisdom that warming is currently occurring with an atmospheric buildup of GHGs, and that the current warming is due to human activity. Natural climate change could be playing a role as well, however. If so, this only reinforces the need for an adequate control response. Thus, a clear alarm to humankind is in order. A major concern of this book is the inadequacy of the range of current mitigation activities to reduce or eliminate human-produced GHGs, although attempts to stabilize and then reduce the levels of atmospheric GHGs are appropriate. A wider range of efforts is needed, thus my call for a Plan B to supplement the ongoing Plan A.

In this chapter, I investigate natural climate change as occurring on two levels: the large-scale changes associated with major fluctuations in the cycles of ice ages, and smaller changes that are my focus. These smaller changes are the climate changes that have occurred within the current interglacial period. That is, those that have affected Earth over the past ten thousand years. I touch briefly on the long-term climatic experience of Earth to provide a broader context, but focus here on the more recent climate within the current interglacial period. Attention is given to the five most recent natural warmings of the last four thousand years, that is, the current interglacial period, for which there is some human record or evidence. Obviously, the body of evidence is greater as we approach the most recent past warmings, those of the Medieval period and the current warming we are experiencing.

The large long-term changes in Earth's climate, such as glacial cycles, appear to be driven by variations in the planet's orbit. The influence of the earth's wobbly spin on its axis affects its tilt toward the sun, and hence changes the way Earth captures the sun's energy; it is thus an important force in climate change. This affects Earth's seasonality and impacts the advance and retreat of glaciers. Changes in Earth's elliptical path also affect its position and distance vis-a-vis the sun.

Orbital drivers of climate were first systematically described by the Serbian astronomer and climatologist Milutin Milankovitch in 1938. Milankovitch was investigating long-term climate change and glacial events such as ice ages, including our current glacial period. He showed that glacial periods closely follow aspects of Earth's orbit and result in cycles, now called Milankovitch Cycles.[5]

The much shorter variations in climate within an interglacial period, however, seem to be due to nonorbital factors. These include natural occurrences on Earth, such as the presence of volcanoes, shifts in ocean circulation, the presence and types of clouding, as well as solar factors like the intensity and timing of heat from the sun.[6] Factors like the earth's distance from the sun are important, but so are cycles in the sun's energy output, which, for example, are responsible for climate changes associated with sunspots,[7] which can contribute to a period of a warmer Earth. Even the position of the continents can be important, since continental positioning can affect ocean currents. However, this is a much longer-term phenomena and typically manifests through tectonics, the shifting of the rock plates that make up Earth's crust. Additionally, of course, the types and prevalence of greenhouse gases, such as carbon dioxide in the atmosphere, will affect Earth's temperature and climate.

RECENT WARMING EXPERIENCE:
THE MEDIEVAL WARMING

The most recent interglacial period provides compelling evidence of climate fluctuations involving both warming and cooling periods. Substantial debates occur regarding the nature of earlier warmings, particularly the Medieval or Viking Warming. That period, which surely predates any significant human-caused influences on warming, saw elevated temperatures beginning with the early Viking trips to Greenland, known among the Nordics as Vineland, around 1000 CE. The climate of Greenland proved so attractive at that time that there was the subsequent establishment of Nordic settlements. Here Nordics established a European lifestyle, including agriculture and the raising of European-type crops, such as hay for the herding animals, and perhaps even grapes, which required a warmer temperate climate no longer found in Greenland. During this same period, there was substantial evidence of a more temperate climate in Europe and evidence that Iceland, which was colonized in the 900s CE, was successfully raising grains.[8]

Although it is the most recent and well-known warming prior to our own time, lessons from the Medieval Warming period are often ignored by contemporary analysts of our current warming. One reason may be that although significant warming occurred, GHGs do not seem to be part of that earlier warming process. Furthermore, analysis of ice cores reveal that the warming was not accompanied by increases in carbon dioxide or GHGs (Figure 2.2), thereby suggesting the warming was driven by a separate and natural climate driver.[9] Thus, even the remote possibility that the human burning of wood—a common occurrence during the period—could release substantial amounts of carbon dioxide are not confirmed in the ice cores. However, the evidence for climate warming over that period is overwhelming, with a thriving European-style society in Greenland for about three hundred years during the Viking period.[10]

This evidence, however, does not fit neatly into the well-known "Hockey Stick" analysis of Michael Mann, a professor of climate science at Penn State. He has made an argument for the unique significance of our current warming by maintaining that both temperature and carbon dioxide levels were stable for a millennium before beginning to rise in the twentieth century. Connecting intertemporal annual data points for temperature over this period, he finds a tight correlation between temperature and carbon dioxide, with a flat line from the fifteenth until the twentieth century, at which time the line moves up gradually at first, and then more steeply, particularly after 1970.[11] I note that this approach follows a relatively clear Hockey Stick shape form of progression—a straight line constant level of temperature and carbon dioxide, indicating a tight correlation followed by a gradual and then sharper rise in both carbon dioxide and temperature.

However, a major issue involves the time period examined. For temperature and carbon, Mann's clear Hockey Stick is formed if the time period for measurement begins about the year 1400 CE. The date chosen by Mann precludes the inclusion of temperature and carbon for the Medieval Warming (950–1350 CE). Here we find that the warming is ending about 1350 CE, with temperatures declining sharply thereafter. However, there is no change in atmospheric carbon dioxide levels. An implication of the breakdown of the Hockey Stick relationship is that carbon dioxide had little or nothing to do with the warming of the Medieval period, and the tight correlation of temperature and carbon was likely simply happenstance. This implies that natural warming, such as the Medieval Warming, can occur without either carbon dioxide or human input. Obviously, a major warming, even as far back as thousand years, makes a current warming appear somewhat less unique.

The Hockey Stick debate is now more of a minor quibble than a fundamental issue. When it is acknowledged that GHGs apparently played no serious part in the Medieval Warming, although they may be driving our current warming, the comparisons between the periods, while interesting, cease to have the same comparative analytical significance. Furthermore, it is now widely accepted that the Medieval Warming was caused by increased solar activity and perhaps even a change in volcanic activity.[12] Other evidence suggests ocean circulation patterns may also have played a role.[13] Finally, this behavior shows the major difference between the Medieval Warming and the current warming is the lack of a carbon role in the earlier warming.[14]

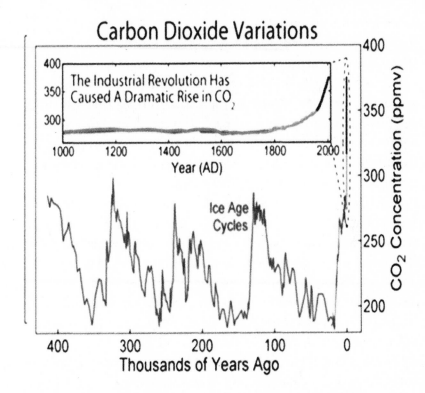

Carbon Dioxide Variations

Figure 2.2. (Image created by Robert A. Rohde, from Global Warming Art, licensed under the GNU Free Documentation License.)

Note: Figure 2.2, replicates the long-term carbon relationship of Figure 2.1, but includes a carbon dioxide graph for the past thousand years.

Nevertheless, there still remain arguments over the similarities and differences between warmings. These involve whether the current temperatures are higher than those of the Medieval period[15] and debates over how geographically pervasive the various warming events were. Was the Medieval Warming confined to a relatively small area of northern Europe or was it a global phenomenon? Some researchers continue to follow the issues raised by the Medieval Warming and argue that the Medieval climate change periods could have been started by an oceanic current cycle that triggered the Medieval Warming climatic changes and could have been transported via ocean currents.[16]

Although there is evidence and arguments for both positions, a broad-scale global warming and a more localized warming,[17] a definitive answer for the precise geographic extent still remains contentious. Substantial historic evidence for an extensive warming effect is provided in the form of known products produced in different areas at those times.[18] For example, early dates for grape harvesting in Switzerland and wine production in Britain are related to climate conditions and are products of times with favorable warmer climates. Additional evidence includes changing historic tree line elevations, which provides evidence as to the level of prevailing temperatures: the warmer the temperatures, the higher up the mountain trees will grow.[19] Other factors used for dating include ancient seeds and insect types found in bogs and elsewhere, tied to particular historical layers of the bog.

Some scholars have gone beyond individual products to attribute broad economic expansion and decline to climate factors related to warming. Thomas Moore of the Stanford Hoover Institution, for example, attributes the increased wealth of Europe, China, India, and even for pre-European North America to the widely favorable global climate conditions of the Medieval Warming time. Moore also notes that the lowering of temperatures at the end of the Medieval Warming period was accompanied by broad agricultural and economic declines across much of the globe.[20]

Similarly, Wallace S. Broecker of Columbia University argues that the Medieval Warming was global and that the present warming we are experiencing should be attributed in part to a current natural warming oscillation upon which the human warming due to GHGs is superimposed.[21] In *The Great Warming*,[22] Brian Fagan also points to the evidence of a favorable multiregional climate warming during this period.

NATURAL CLIMATE CHANGE

Debates continue as to the nature of the current global warming and its long-term implications for humankind. However, as the above demonstrates, climate change is not unique to our time or even our interglacial period. There is no evidence that humans had a role in earlier warmings. Natural climate change has occurred before, and indeed appears to be essentially continuous.[23]

However, two climate elements appear to be unique to our current warming. First is the role of carbon dioxide and other GHGs in driving warming. Second is the role of humans as users of fossil-fuel energy, and thus as agents that are responsible for the release of GHGs into the atmosphere.

As I argue above, it is unlikely humans had any significant part in any of the earlier warmings. Humans might have been involved in deforestation in a minor way, as they had the use of fire. But fire had been releasing carbon from vegetation through natural processes, such as biological growth, decomposition, and wildfire, for many millennia. Indeed, the carbon release fires over the Russian steppe and the Canadian North, started by frequent lightning strikes, would dwarf the fires started by humans. In any event, the level of carbon in the atmosphere remained constant over the period.

The impact of the carbon released was evidently small,[24] since GHGs did not increase (Figure 2.2) and were apparently not drivers in natural earlier warmings. Additionally, ice cores and other evidence do not show a significant increase in atmospheric carbon, as reflected in the thousand year atmospheric carbon trend shown in Figure 2.2. Although the current warming is at least partially due to the buildup of GHGs, this is clearly not the case for the Medieval Warming, which must have been due to other forces.

Following on the above point, a unique feature is the role that humans are playing in the current rapid growth in the level of atmospheric carbon dioxide, primarily through fossil-fuel energy. Despite wildfires that had for ages swept across the forests, steppes, and plains of Eurasia and North America, the aggregate net carbon changes were small to negligible,[25] as reflected in the absence of any rise in the estimated carbon dioxide in the atmosphere.

The conclusion must be that, whatever the cause, the Medieval Warming was not the result of a rise in the atmospheric level of carbon dioxide. So, the major drivers of the two most recent warming periods must essentially differ.

THE ROLE OF CARBON

In his book, *The Global Carbon Cycle*, David Archer, a well-known University of Chicago geoscience professor, estimates that atmospheric carbon levels have remained essentially contained below 270 for eons (Figures 2.1 and 2.2).[26]

Figure 2.1 shows the relation between carbon and warming over long periods of time—hundreds of thousands of years—much longer than our interglacial period. According to this figure, carbon dioxide is correlated with warming over very long periods, namely tens of thousands of years, but seems to be less so over shorter periods, of centuries and millennia. Figure 2.1 includes the carbon relation over the past one thousand years. Consistent with this behavior, it is clear from ice core samples that atmospheric carbon dioxide and GHGs were relatively low in the early part of our current interglacial period, yet warming still did occur.

At the end of the most recent interglacial period before our own, some 135,000 years ago (see Figure 2.1), the carbon dioxide (CO_2) level in the atmosphere was about 270 parts per million (p/m). This is a level that it would not reach again until the year 1750 CE in our own interglacial period.[27]

Although carbon evidently did not play a major role in the Medieval Warming, or indeed in any of the other interglacial warmings of the past ten thousand years, carbon appears to be important both for long-term climate, and particularly for our current climate change and warming. David Archer examines the global carbon cycle and its relation to climate warming. He points out that a natural carbon cycle involves a continuous shifting of carbon, often involving very long time periods, perhaps hundreds of thousands of years, among the various carbon sinks, e.g., the atmosphere, the oceans, and the terrestrial system, which includes biological systems, forests and plants, as well as soils and rock materials. Large amounts of carbon are captured as various compounds in rock and land formations, and some of these will eventually be released through weathering over very long time periods.

Archer stresses the uncertainty of warming when he asserts that "the natural carbon cycle is a wild card . . ." He goes on to state, "we are unable to predict how the ocean carbon cycle will eventually respond to the provocation of our (human) fossil fuel release." The release of the ocean's carbon would have implications for the level of carbon in the atmosphere and hence the degree of warming. Uncertainties abound regarding how human behavior may inadvertently affect the carbon cycle through and including its direct contribution to GHGs and on the natural carbon system. Archer cautions that "science still knows very little" about Earth's natural carbon cycle and the role of cyclical subsystems, e.g., the ocean carbon cycle, within that larger system.

Finally, note that the discussion above provides a context for readers to assess the likelihood of climate change driven by natural systems as well as human. The human-generated levels of carbon in these periods were undoubtedly far below the volumes from burning that occurred after the beginning of the Industrial Revolution and particularly with the advent of the widespread use of fossil fuels.

THE IPCC AND SOURCES OF
NATURAL CLIMATE VARIATION

Science has yet to definitively identify a non-GHG natural driver that could account for many nonhuman-generated warmings. Indeed, when assessing the completeness of the Intergovernmental Panel on Climate Change (IPCC)'s explanation of global warming, climatologists Wigley and Santer[28] conclude that human climate influences were responsible for 50 to 150 percent of observed warming from 1950 to 2005. Notice that this range is consistent with both the notion that humans were the sole factor and that human impact was accompanied by a significant natural warming component. The IPCC, the UN organization responsible for monitoring warming, generally acknowledges that a portion of the warming is driven by other nonhuman GHG factors. It states that "more than half of the observed increase in global average surface temperature from 1951 to 2010 was caused by the anthropogenic increase in greenhouse gas concentrations."[29] By noting how much warming can be accounted for by human activity, the analysis reveals that up to 50 percent of current climate change is not explained by the current human-driven carbon/GHG theory. This void provides recognition that something else—I call it "natural climate variability"—is occurring simultaneously, and this something must be some sort of natural driver or set of natural drivers. A gap in our knowledge is clearly present, and it remains for science to explain.

Some of the explanation may be found in the sun. It is well-known that the sunspot cycle does effect on the earth's climate by slightly raising and reducing temperature over an eleven-year cyclical period. The sun's increased radiation is one of the more common explanations of the Medieval Warming.[30] Although the IPCC has disqualified present-day solar activity as having much of an impact on current long-term warming, it has not yet found a mechanism currently active by which the warming is accomplished. Some potential solar drivers are being hypothesized by scientists and researched in the literature, as discussed below.

An appropriate question being investigated in some scientific circles is: what forces could have generated the climate change revealed during the early interglacial climate change periods, and are these forces active today?

POSSIBLE SOURCES OF NATURAL CLIMATE VARIATION

Since carbon change is not always a part of a global warming, other nonhuman natural factors may have been the drivers of the climate warming through a mechanism other than carbon dioxide. These could include solar changes, volcanic activities, and other forces discussed below. Note that the logical implication of a cooling period, about which there is no doubt, is that it would have to occur adjacent to warmer periods, almost by definition.

There are a number of candidates for the unrecognized climate driver. The most likely is probably some source of solar energy. Solar energy is not currently viewed as a major contributor to today's warming by the IPCC. However, solar factors are still not yet well understood. When discussing a new solar space probe named in his honor, astrophysicist Eugene Parker stated, "We have to be prepared for some surprises—things we never thought of, or things that we thought of but were not correct."[31]

European scientists in particular have been tenacious in their research into pathways whereby the sun may be impacting on Earth's climate. Studies, now widely accepted, suggest that the sharp and continuing cooling was likely due to a measured decrease in solar variability at the beginning of the 1400s. During the Little Ice Age, there was a minimum in sunspots, indicating an inactive and possibly cooler sun. This absence of sunspots is a period called the Maunder Minimum, which occurred during the coldest period of the Little Ice Age between 1645 and 1715 CE, when the number of sunspots was very low. This cold period is named after British astronomer E. W. Maunder, who discovered the dearth of sunspots during that period and drew climate implications. The lack of sunspots meant that solar radiation was probably lower at this time, and this cooling lasted into the early 1800s.[32] Could it be that the current warming is simply a rebound from that multicentury cold period, as suggested by some researchers?[33]

Recent work by Swiss researchers, including Werner Schmutz, a researcher at the Swiss Federal Institute, demonstrates that an assumption of stronger solar fluctuations can explain much of both the past and current climate, both warmer and cooler.[34] Based on this consideration and findings of later related work, Schmutz found that the end of the Little Ice Age could be due to a world that has been experiencing solar-driven warming influences since the late 1800s.[35] This interpretation is consistent with the end of the Maunder Minimum. The timing also coincides closely with the onset of the Industrial Revolution and the beginning of significant human fossil fuel GHG emissions, initially from the burning of coal.[36]

The role of solar variability in the Medieval Warming may be even more likely given the experience with the Little Ice Age in the intervening period between the Medieval Warming and the 1800s.[37] Although models and temperature reconstructions suggest this would have been enough to reduced average global temperatures, it would not be enough to explain the entire magnitude of the regional cooling of the climate in Europe and North America.[38]

CLIMATE MODELS

The rationale for the focus on GHGs as the source of the current climate warming is found in the use of climate models. It is maintained that models with only natural inputs do a fairly good job of projecting the recent past up to about the year 1970. However, beyond that time, the models perform poorly in predicting warming unless the human impact with the introduction of rising carbon dioxide and GHGs is added.[39] Hence, this is viewed by many as the "smoking gun" that provides the critical evidence for a GHG-driven warming.[40]

The Fifth IPCC Global Warming Assessment in 2014 recognizes solar activity as a potential climate change agent. Solar activity can change over time, increasing or decreasing the heat brought into Earth's system, thereby changing climate. Climate models assume different magnitudes or values for the climate variables as they undertake sensitivity analyses. Researchers have noted that if a larger role in climate modeling had been given to solar effects, the association between the driving force of GHGs and warming would be diminished. This is not as simplistic as it seems.

Note here that climate models do not always have definitive real-world data of past levels of solar radiation fluctuations, but rather must draw on reconstructions based on other evidence and reasonable assumptions to establish their variable and parameter assumptions for earlier periods. As with much of their past data, they must sometimes draw from a reasonable range of best guesses or parameter assumptions. Thus the results are based on modeler discretion as well as the model itself.

Kelvin Droegemeier, President Trump's top science and technology adviser, repeatedly points out the limits of climate models. "The evidence of the models suggests that it (climate change) is human induced or there's a strong human signal . . . but we don't know everything there is to know . . . about the carbon cycling. Carbon sequestration. We don't know."[41]

Undeniably, there are a large number of different climate models that have been used to estimate a range of future temperature outcomes.[42] Cess[43] used fifteen different models twenty-five years ago in his comprehensive assessment of CO_2 radiation forcing (that is, the flow of energy into and out of the climate system), while Collins used twenty updated models more recently to re-examine forcing.[44] In both cases. the radiative forcing differs substantially among the different models leading to differences in model forecasts.[45] Thus, although the shape of the Mann's Hockey Stick captures the temperature/carbon correlation and trends since 1400 CE, including the Little Ice Age, it totally misses the phenomenon of the Medieval Warming of 950 to 1350 CE. In short, models can be useful but can also lead astray.

RECENT RESEARCH ON SOLAR SOURCES OF NATURAL CLIMATE CHANGE

Science accepts the concept that certain natural activities might generate climatic change under some circumstances. These include solar activity, volcanic activity, ocean currents, and plate tectonics, as well as certain types of clouding. All of these forces have the potential to affect warming. The most intriguing of these is probably the potential tie to solar activity, given the continuing role of solar activities, and changes in these in the earth's climate.

A number of scientific groups have been examining how solar activity might be a climate change driving agent. In addition to the Swiss, very recent research groups in Sweden and particularly Denmark[46] have found evidence that solar activity could currently be forcing climate change and temperature activity. They see this change as being potentially significant. The compelling feature is that natural solar forces may be contributing to the current warming in a manner not yet fully understood.

Recent work demonstrates that an assumption of stronger solar fluctuations can help explain both current and past climate.[47] As noted above, Werner Schmutz found that the end of the Little Ice Age could have been caused by the solar-driven warming influences since the late 1800s.[48] Other recent research also challenges the IPCC's assumption that recent solar activity is insignificant for climate change and that the same will apply in its projections for the future. For example, a study in 2014 published in *Nature Geoscience* by Florian Adolphi and Raimund Muscheler of Sweden's Lund University used new data that shows a "persistent link between solar activity . . . and climate change."[49]

Similarly, a team of Danish researchers lead by Henrik Svensmark, of the Division of Danish Solar System Physics at the Danish National Space Institute, has examined evidence that variations in the energy emitted by the sun could be a significant climate driver. Their basic premise is that clouding is important to climate, and the sun's cosmic rays affect the earth's clouding. Clouding in turn affects the earth's temperatures.[50] Although often criticized, this research continues to create significant interest. The link is that solar cosmic rays create atmosphere aerosol, which in turn contributes to clouding. This is particularly important since clouding is widely recognized among climate modelers as both likely to be an important determinant of the earth's climate but also one of the most poorly specified variables in climate models.

Modelers acknowledge that clouds are important to climate change and warming and yet admit that their role is neither well understood nor well represented in climate models. For example, V. Ramanathan, a climate modeler and professor of atmospheric and climate sciences at Scripps Institute of Oceanography in San Diego, notes that while climate models have become a basic tool for climate researchers, "when it comes to clouds they (climate modelers) have a long way to go."[51] Although still hypotheses, the research noted above reflects a continuing acknowledgment of the incompleteness of our current understanding of the influence of solar forces.

It is also worth noting that discussions of geoengineering, i.e., engineering approaches to dealing with climate change (see Chapter 5), often focus on technologies that would affect climate change by changing global reflectivity by whitening clouds, thereby reducing the earth's capacity to capture the sun's energy. Whiter clouds, it is hypothesized, should reflect more of the sun's heat, thereby offsetting some of the warming associated with increased GHGs. Additionally, recent experiments by the European Organization for Nuclear Research (CERN), using particle accelerators and detectors, has given increased credibility to a solar driver, which promotes cosmic ions and thereby affect earth's clouding.

These findings all suggest that science's knowledge of the role of solar forces in climate is still incomplete and could be critically important to a more complete understanding of climate change. This research suggests that natural mechanisms like the sun explain a significant portion of climate change in the past. Can it be that the climate changes that the IPCC notes cannot be explained by human-induced actions may need to be explained by natural implementing mechanisms? And, if so, does this have implications for human attempts to survive future warming?

As suggested by some of the ongoing research, the driving force may be solar but implemented in different ways than understood up to this time. Rather than speculate further, my study is focused on the simple GHGs global warming model while allowing room for natural forces. As such, I question the adequacy of the current mitigation approach to addressing warming, while suggesting the usefulness of a supplemental adaptation approach.

OTHER NON-SOLAR MECHANISMS

Climate change also may be influenced by natural forces other than the sun. We know that ocean cycles, such as El Niño Southern Oscillation, involve a cycle of warming and cold sea surface temperatures.[52] These can have large-scale impacts on both ocean processes and on global weather and climate. Some researchers have suggested that more fundamental and pervasive climate change could have begun with an oceanic current cycle that triggered the Medieval Warming by redirecting tropical heat.[53] These localized climatic changes, in turn, could have been transported globally via ocean currents. Finally, volcanoes and plate tectonics are recognized to influence the earth's climate through gas and particle emissions into the atmosphere from volcanoes, or long-term land relocation from tectonics. These natural forces all have the potential to contribute to climate change and are consistent with the hypothesis that more than one force may be driving current climate warming.

Although some of the emerging scientific work has been controversial, continuing research activities by prominent scientists constitute an acknowledgment that the carbon/GHG approach is viewed by many as only the beginning of a broader understanding of the entire climate change process. While many of these research findings are preliminary and not conclusive, they indicate that our current understanding of climatic change is still in its infancy and deficient as a complete scientific explanation.

HISTORIC WARMINGS IN POST-GLACIAL PERIODS

Humans, as a species, have populated Earth for at least several hundred thousand years. But it is only in the last five thousand years or so, the period for which we have written records, that human societies might be properly characterized as advanced. It is at the beginning of this period that the first cities were being established in Mesopotamia and humans began to move away from traditional tribal, often nomadic social systems.

Scientists have identified a modest number of climate warmings since the beginning of our latest interglacial period, perhaps eight in the past eight thousand years (see Figure 2.3 and Table 2.1). These figures show temperature variations since the last ice age. The temperature variations are modest, ranging roughly around plus or minus 0.5 degrees Celsius. The current warming has been roughly 0.8 degrees Celsius, although less on the graph, but has been within the "historic" range.

Only five of these warmings have been in the human historical period (see Table 2.1). Two were in the Bronze Age, perhaps around 1700–1500 BCE and 1200 BCE, respectively. The first of these might have been associated with the movement of Joseph's family into Egypt to flee the persistent famine recorded in the Bible.[54] The latter of these climate changes appears to have generated drought in the eastern Mediterranean and Asia Minor and is associate with massive crop failures. This climatic event may have precipitated the collapse of the Bronze Age circa 1200 BCE by creating the mass movements of peoples that led to widespread disruptions and warfare in the eastern Mediterranean, and the influx of new peoples including the Philistines. It is perhaps associated with the creation of the political state of the Jews.[55]

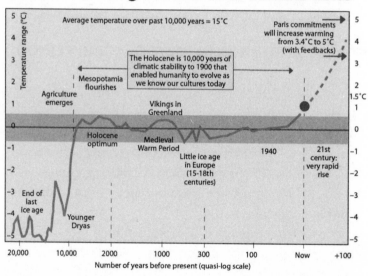

Figure 2.3. Temperatures for the past two thousand years. Projections for the next one hundred. (Chart courtesy of David Spratt.)

Figure 2.3 traces data related to warming (thick line) in the post-glacial period. It highlights particular warming events, including the Medieval period, one thousand years ago, the Little Ice Age, and the current warming. It also projects a continuation of warming through the twenty-first century.

Table 2.1. Historical Warming Periods	
End of last ice age	10,000 BCE
Minoan warming	1700 BCE–1500 BCE
Bronze Age Collapse	1200–1000 BCD
Roman warming	200 BCE–400 CE
Medieval warming	850 CE–1320 CE
Current	1830 CE–present

(Table from Fred Singer and Dennis Avery, *Unstoppable Hot Air: Global Warming Every 1500 Years* [Lanham, MD: Rowman & Littlefield, 2006], as modified slightly by the author.)

Over one thousand years after the collapse of the Bronze Age, the fall of Rome began in the third century CE. According to Kyle Harper of the University of Oklahoma,[56] cooling and drought descended, and rains began coming late to much of the empire. This period in the eastern Mediterranean was quite different climatically from later and current periods. For example, in that time, the Egyptian city of Alexandria had rain every month but August. Today, it is in an arid climatic zone in which months go by without rain. There is evidence that vines and olives were cultivated farther north than today, suggesting a warmer climate in that period. By the fifth century in Rome, the drought had become widespread and had undermined Roman agriculture and provoked mass migrations that ultimately contributed to the fall of the empire.[57] One half-millennium later, after a cooler period, the Medieval Warming came about, lasting roughly from 950 CE to 1350 CE. This was followed by a sharp cooling period called the Maunder Minimum, leading into the Little Ice Age, which continued into the 1800s.

Of course, all but the most recent of these periods are some considerable distance back in time as human history is measured, and the information we have is mostly that of physical remains. In additional to tree rings and seed remains, these importantly include ice or seafloor cores, which are long cylindrical cores drawn from ice formations or the seafloor. These cores often contain air pockets from earlier periods, which can be investigated for carbon dioxide levels and other climate-related data. The air pockets are small but numerous and thus the findings can be confirmed multiple times. These sources provide information related to temperatures, carbon dioxide levels, and precipitation, among other relevant indicators.

The period since the beginning of the Medieval Warming (including our own period), however, is obviously much richer in information and useful data. A comparison of the Medieval Warming period and current warming periods may offer a useful perspective on both. The conventional wisdom is that these earlier warming events can be attributed to increased solar intensity or reduced volcanic activity on Earth during that period,[58] which often persisted for up to several hundreds of years, as may have been the case in the Medieval Warming.

It is recognized that the evidence for these warmings becomes more tenuous as their distance in time from us increases. Nevertheless, evidence of the occurrence of the several historical warmings indicates that they are more frequent than commonly believed.[59] Again, there is no evidence of serious systemic GHG fluctuations until our current period (Figure 2.2).[60]

As previously mentioned, the following cooler period, from 1400 CE to the mid/late 1800s, is sometimes known as the Little Ice Age. Research confirms that temperatures fell precipitously in the 1400s, before which time the remaining European colonists apparently left Greenland permanently. The initial cooler period was followed by the Maunder Minimum, which is associated with the Little Ice Age. The Little Ice Age climate cooling "trough" was punctuated by short, extremely cold periods, followed by modest temperate recoveries, with the temperature fall creating overall a multiple-century cooling period lasting into the 1800s.[61]

From the latter part of this period come anecdotes regarding the colder weather in Europe and North America.[62] These include activities such as ice skating on the Dutch canals, a frozen river Thames, poor agricultural production across Europe, as well as George Washington's famous crossing of the Delaware River through the ice flows during America's Revolutionary War.[63] These colder weather conditions no longer occur and have not been seen in modern times (which is evidence of a warming since the Little Ice Age). This period was followed by gradual warming driven by a buildup of carbon dioxide and GHGs, presumably associated first with human-created coal-powered Industrial Revolution activities, and subsequently by persistent increases in fossil use leading to our current era.

Another interpretation of this pattern discussed earlier and noted by Schmutz is that perhaps we are not currently experiencing a unique warming but rather a simple return to a historic normal that has been suppressed in the post-Medieval Warming period by the cooler Little Ice Age period.[64] Although quite different conceptually, these two views need not generate empirically different outcomes.

The occurrence of the Medieval Warming period provides a nice contrast to the current warming period. What are the differences and similarities? In scientific terms, these two cases could be viewed as providing what is sometimes called a control group and an experimental group. We have a control group, the Medieval or Viking Warming, which has been fixed in one environment with a unique set of outcomes. These results can be compared to today's experimental group, which is being subjected to increasing human-generated GHGs and associated warming. For one of the two similar warming periods, the Medieval Warming, we find no GHG increases and few if any human drivers. For the current period, by contrast, there is a definite buildup of human-driven carbon and other GHGs. Although it is sometimes argued that the temperature is warmer now than in the Medieval period, it is clear that the temperature of the Medieval Warming was significantly warm by long-term standards. Again, the warmth of that period suggests a role for natural warming forces that is not based on GHGs.[65]

Many want to treat today's current warming as unique and perhaps it is in the sense that it is clearly the only warming driven by human-generated carbon dioxide and GHGs. This situation makes the human challenge of surviving a global climate change of their own making even greater.

THE GRAPHICAL INTERGLACIAL RECORD: CARBON DIOXIDE AND TEMPERATURE

For the separate shorter time period graph in Figure 2.2, note that the carbon dioxide level is constant for the past thousand years, except for the very recent period. The most interesting feature is the absence of any change in atmospheric carbon dioxide levels despite the increase in temperature during the Medieval Warming. This finding, of course, challenges the notion that high levels of carbon/GHG are the drivers of warming. High carbon/GHG levels apparently were not necessary for the Medieval Warming, so other natural factors were at work. As noted earlier, in addition to possible solar influences, these may have included forces like powerful ocean currents that brought tropical waters to warm the northern regions.[66]

Figure 2.3 finds similar results to the trends of earlier figures. Again, the global temperature stabilizes in the period between eight thousand and ten thousand years ago, with the subsequent variations being consistent with earlier prehistoric climate warming levels and the historic climate warmings going back prior to the Bronze Age, beginning circa 3000 BCE. Note that the temperature increases are of roughly the same magnitude as those experienced earlier.

I should note that I do not want to argue too strongly for any one figure or any one point on a graph. The important point here is to recognize that the basic trends are very similar across graphs. This is true even when different researchers construct the graphs.[67] A warming point on one graph is also found on the others. Further, warming periods on the graphs that are generated by physical evidence are found to occur at the same time as the periods discerned separately by historic means. For example, records using glacial estimates of temperature are generally consistent with warm periods identified by cropping types and tree rings not only in Greenland, but throughout Europe.

SUMMARY OF NATURAL CLIMATE CHANGE AND THE WARMING EVIDENCE

Although most of the literature on warming focuses on the role of carbon and GHGs, there is compelling evidence that other important forces exist. Over the long-term, Milankovitch Cycles (that is, the orbital changes of Earth's relationship to the sun) explain reasonably well the comings and goings of major glacial periods. For shorter warming cycles during interglacial periods, solar impacts as well as a number of planet-centric explanations have been discussed. As we have seen, the evidence for a closer event in time, the Medieval Warming, is more substantial.

Let us now review the comparative evidence for the current warming period. First, the historic record provides bountiful evidence of warming periods unrelated to carbon dioxide or human activities. It is now widely agreed that there is indisputable evidence of a Medieval Warming for several centuries around one thousand years ago. There is fainter evidence of earlier warmings in our interglacial period both during historic times (when written records were available) and prehistoric times. The more distant evidence uses ice core, tree location and growth, seed locations, and other data to provide estimates of temperature, precipitation, and vegetation in the earlier periods. So, the natural system periodically generates warming (and cooling) during the interglacial phase.

Second, the Intergovernmental Panel on Climate Change (IPCC) acknowledges that it cannot explain all of the recent warming after 1951; rather, it can only explain something over one-half of it. If their assessment is correct, there is more than the single human force influencing climate.

Third, some recent research, although still preliminary and tentative, ties natural solar phenomena, such as solar cosmic ray impacts on clouding, to short-term warming.[68] The demonstrated potential of non-GHG forces and the inability to fully explain current warmings with a narrow carbon/GHG hypothesis gives credibility to the existence of other natural climate changing forces. These forces give support for existing, although not yet well understood, nonhuman forces that may affect short-term climate.

CLIMATE CHANGE AND THE PUBLIC

The common way of approaching perceived evaluations and estimates of climate change for public consumption is to treat Earth's basic climate system as being in a comparatively steady equilibrium and then look for deviations from that steady state that we can attribute to human-caused change. An example is found in Al Gore's book *An Inconvenient Truth*, which treats deviations from a steady state as solely driven by the buildup of atmospheric GHGs. This approach, which ignores any natural climate variability, is also articulated extensively by the media as well as in most broad discussions of climate policy.[69]

Similarly, climate forecasts of anticipated changes are typically based on assumed climate changes due to human causes in an underlying essentially stable climate system. What is missing, of course, is the recognition and incorporation of a role for natural climate change that can on its own destabilize the climate system, irrespective of what we humans might do to affect it. An example would be the volcanic eruptions in recent years that have spewed large amounts of uncontrollable smoke and ash into the atmosphere, as well as the noxious gases produced by the Kilauea volcano's lava meeting the ocean in Hawaii. More broadly, the absence of a public discussion of the role of natural climate variability skews the entire discussion of the selection of policies that deal with both human and natural climate change. This becomes increasingly important when the discussion approaches global policy issues such as the Paris Climate Agreement or US contributions to a Global Climate Fund. These organizations have developed programs, based predominantly on mitigation, to address climate change.

Despite compelling evidence of the occurrence of periodic climate change in the post Ice Age period and the acknowledgment of an incomplete explanation offered by the GHG approach of realized climate change events, the insistence on focusing the analysis and the discussion almost entirely on the human drivers not only misinforms the public but is could well result in wrongheaded and inadequate policy responses.

MITIGATION IMPLICATIONS: GHGs ARE NOT THE WHOLE ANSWER

It has been demonstrated that substantial climate variability has been generated in the current interglacial period via the natural system. Natural forces have generated several warmings, even in historic times. The overwhelming majority of international climate scientists appear to accept the view that the current warming is unprecedented and human caused. However, as I have demonstrated, there are likely to be two major forces, natural and human, that could be currently driving climate change. When two independent factors are driving a system, an analysis focusing exclusively on one of these factors may provide an inadequate examination of the scientific phenomenon. That is, GHGs are not the whole answer.

The recognition that there are both human and natural forces that can drive climate change is critically important to the development of a comprehensive understanding of the climate change process. We now have a positive mechanism to explain the multiple forces driving a significant portion of historic and currently observed climate change. This broader perspective allows for the development of a more comprehensive and robust policy regime.

CONCLUSIONS

A world where warming disturbances are driven by natural phenomena may call for a different type of human response than one where warming is a by-product of human activity such as energy production. Although a series of Earth warmings has been driven by non-GHG forces, and non-GHG factors may be involved in the current global warming, the case for claiming that GHGs are the driving force of the current warming is quite compelling. A look at the climate record during the interglacial period shows that over decades and centuries temperatures regularly cycle from warmer to cooler and back, although in a fairly modest range. These changes appear to have little to do with atmospheric carbon dioxide levels. Thus, controlling GHGs is no guarantee of climate stability.

I believe that society has initially made a correct decision to address warming by focusing on mitigation of carbon dioxide and GHGs, which appears to be the *dominant* warming driver. However, over the longer term, the objective may no longer be to simply prevent or mitigate a known human-generated problem. In the case where the issue is carbon/GHG emissions as the cause of warming, some process to control the source of these emissions, fossil-fuel energy, and their climate impact is appropriate. If warming is also part of a recurring natural process, however, it will require a human population capable of dealing with that source, even though humans may not have created it and cannot directly control it. The problem now becomes one of addressing warming forces that are separate and independent of human activity.

Finally, while I believe that the current approach of spending most of our human energy and resources on mitigating (preventing) human climate change has been sensible in the past, I also believe that to focus exclusively on that approach is now is shortsighted and inefficient. If natural climate change is occurring together with human-generated change, prevention or mitigation efforts alone will be limited in their positive effects. Mitigation cannot prevent the damage due to the part of warming generated by the natural system. However, even if natural climate warming is not a current problem, I believe a less risky and more productive approach is to manage (adapt to) the damage generated by climate change from any source while continuing an aggressive GHG mitigation policy. Hence, the challenge becomes one not only of mitigating human-driven warming, Plan A, but also that of adaptation and damage control, Plan B.

CHAPTER 3

PLAN A:
MITIGATION—
A BRIDGE TOO FAR?

I n 1944, the European Allies devised a high-risk plan intended to end the war that year. By moving quickly through Holland, the Allies hoped to bypass major Nazi defenses and quickly defeat Germany. The plan required an airborne attack to capture a critical but distant bridge. The plan turned out to be too ambitious—a bridge too far. Many lives were wasted and the war ground on well into 1945. The evidence is growing that the mitigation approach to climate change, like the military plan of 1944, is inadequate to its task of fully addressing global warming.

In March 2018, the environmental news website Greenwire had the headline "World unlikely to meet temperature goal—leaked U.N. report."[1] The leak referred to was a draft report being developed in the IPCC as their Summary for Policy Makers (SPM) for the forthcoming Sixth Assessment Report, due in 2019. The draft states that without a rapid phaseout of fossil fuels, there is a "very high risk" that warming will exceed the 1.5 degrees Celsius target. The report went on to note that this could have catastrophic and irreparable consequences for the earth. Further, it states that even if warming is controlled but still exceeds a 2 degree increase, the potential risks to economic development grow dramatically. These very real concerns from within the IPCC raise the question of whether the targets are realistic and within the realm of feasibility, given the approaches that are being used.

There is pervasive skepticism regarding the feasibility of the targets of the Paris Agreement. The Paris Agreement calls for pledges to be carried out through 2030 and then maintained at that level through 2100. In essence, the Paris target is to keep global warming well below 2 degrees, but in any event to pursue stringent efforts to keep warming under 1.5 degrees. The IPCC has stated that 2 degrees is not a safe target but "better seen as an upper limit." Temperatures rising above that upper limit are seen as having a significant risk of "runaway climate change."

Other analysis external to the IPCC raises the same concerns. For example, a recent study by Shell Oil[2] found that the Paris targets were very difficult to meet. To achieve Paris goals, Shell's analysis requires a new technology that can "suck" massive amounts of carbon from the atmosphere. New carbon-sucking technologies would be required to capture about one-half of annual emissions by 2070. Similarly, an MIT study finds that the Paris targets "would lead to a wide range of projected climate impacts around the world."[3] They argue for an alternative and more stringent approach. Michael Oppenheimer of Princeton[4] estimates the probability of the Paris Agreement achieving the climate targets at about 10 percent. Note that this estimate is based on the assumption that the GHG reduction targets are achieved, something that is already in doubt. Most assessments find that current emission reduction performance is not meeting existing Paris pledges.

In short, the possibility of reducing fossil-fuel energy enough to meet the Paris targets appears very unlikely, especially given the energy demands of emerging, developing economies. The potential for the massive capture of carbon with as-yet-to-be-invented carbon-sucking machines is highly problematic. Clearly, a Plan B needs to be developed to supplement the traditional mitigation approach, which is now obviously inadequate.

APPROACHES: MITIGATION VS. ADAPTATION

Through the late twentieth century, fossil fuels were the world's dominant source of energy. Fossil fuels constituted over 80 percent of the earth's energy during this time, and that figure remains almost the same today, although there have been some substantial changes in the composition of fossil fuels away from coal to natural gas, particularly in the United States and some of the developed world. But during this period, demand for energy and fossil fuels has increased greatly. So, despite the fact that there have been significant efforts to wean societies away from fossil fuels, overall the energy mix has remained stubbornly stable for over a half of a century.

Note that during this period the international community has made a host of commitments to reduce emissions of GHGs and the use of fossil fuels, yet societies are still overwhelmingly dependent on them. This performance suggests that the task of stabilizing GHG emission is not an easy one. Daniel Schrag, of Harvard's solar engineering group, estimates that gaining real control of the current mitigation approach will probably take more than a century and cost many of trillions of dollars.[5] This raises the question as to whether our current mitigation approach can, in fact, achieve the desired ends. Developing countries want to be able to use fossil fuels to help them reach the economic levels that the developed nations have enjoyed, even as they share concerns about warming. Actual commitment to international goals is quite uneven, as seen in the variation of countries' achievements of their individual mitigation goals.

This chapter discusses what measures might be undertaken to stabilize or further reduce GHG emissions. The reduction of the use of fossil fuels globally would necessitate the identification and use of alternative energy sources to replace them. In recent decades, the dominant use of fossil fuels with their release of GHGs was supplemented by other energy sources including nuclear power, hydropower, biofuel, wind power, and solar power. Although these alternative energy sources have little GHG associated with their production of energy, their potential as major replacements for fossil fuels is probably limited. Given the limits on nuclear and hydro, the candidates as replacements seem to be predominantly renewables such as wind, solar, and perhaps biogenic power derived from plant materials, such as ethanol and wood.

Approaches for addressing climate change can be included in one of two categories. The first is mitigation, that is, efforts directed to preventing warming directly. The second approach, adaptation, would be to allow the event that causes warming, the release of GHGs, and then to respond. The response would be to try to adapt to the warming event by anticipating what the event might be, and by anticipating and controlling the likely damage through adaptation strategies and actions. Plan B includes adaptations that would focus on neutralizing the damage, as well as anticipating ways to inhibit the extent of the damage.

Adaptation is much like addressing a natural disaster—we may know that a hurricane is coming, but we cannot stop it. However, we can do things to prepare for it, like protecting the windows, moving the car to a safer location, and obtaining supplies (flashlights, water, and food) that will allow us to survive it. If this appears inadequate, we can evacuate and move to a safer location out of harm's way. If more time is available, barriers, perhaps only temporary sandbag barriers, can be strategically placed to protect property. Over longer periods, older infrastructure can be replaced with new more appropriate structures. All of these types of activities can reduce the damage that might be associated with a hurricane, even if the hurricane itself is not preventable. Essentially, these are aspects of an adaptive approach.

Another component of the adaptive strategy involves attempts to neutralize the warming effects of the GHGs. Just as a biologist might try to address an infestation by spraying a liquid that sterilizes insects and thus prevents an expanding infestation from overwhelming a forest, so, too, humans may try to neutralize the warming effects of GHGs. Different approaches might be employed, some involving changing the reflectivity of the atmosphere or the land surface. These and other forms of geoengineering are discussed in a following chapter.

While methane, nitrogenous oxide, and other gases, including hydrofluorocarbons (HFCs), have warming qualities, the dominant GHG is carbon dioxide. Although not the strongest warming gas by individual warming per molecule, carbon dioxide is responsible for the majority of the warming generated by human activity by virtue of its prevalence in the atmosphere. Many people who are not familiar with the gas do not realize that carbon dioxide is not the dirty, polluting gas that we sometimes see bellowing from industrial smoke stacks. In fact, the recent Clean Air Act legislation as amended in 1990 did not name carbon dioxide as one of the individual pollutants to be regulated. In legislation, it was argued that carbon dioxide was not a pollutant in the traditional sense. It is a colorless, odorless, tasteless gas that is visually unnoticed, occurs naturally in the atmosphere, and is necessary for life (plants consume it and release oxygen). It is nontoxic, safe for humans and animals. In the dispute and litigation that followed the passing of the legislation, a US Supreme Court decision determined that even though carbon dioxide was not identified in the legislation, it should be treated as a polluting gas in need of regulation. This was due to its polluting environmental effects with respect to climate change and warming.[6]

PLAN A: SOME APPROACHES TO MITIGATION

A logical starting point for examining approaches to mitigation of warming would be at the GHG source. The GHGs are mostly carbon dioxide released by the human burning of fossil fuels for energy. In the case of human-generated carbon and other GHGs, the basic approach to mitigation is largely the attempt to reduce the flow of human-generated carbon and other GHG emissions into the atmosphere. The major opportunity to reduce or even eliminate this gas is almost certainly at the energy source. The focus would be on reducing the amount of fossil fuel used to produce energy. This is the approach that is overwhelmingly being used today. It is well-known that most human carbon/GHG emissions are the result of the burning of fossil fuel—coal, petroleum, and natural gas—for energy. The burning process releases the carbon from fossil fuels and allows it to combine with oxygen in the atmosphere to produce carbon dioxide.

Through the late twentieth century and into the twenty-first, fossil fuels were the world's dominant source of energy. For energy production to expand, as it will undoubtedly need to do as the world's economies develop through this century and beyond, the reduction of the use of fossil fuels for energy would necessitate the identification and use of an alternative energy source as an energy replacement. The favored candidates to replace fossil fuels are renewables, specifically wind and solar, with biomass a distant third. Renewables gain in attractiveness given the apparent limits on the use of nuclear and hydropower alternatives.

TOOLS TO PROMOTE ENERGY SOURCE CHANGES

A global shift from fossil-fuel energy to renewables use is not an easy task. Fossil fuels have achieved widespread acceptance due to their high energy content, their relative ease of use, and their low cost of production and transport. Additionally, some fossil-fuel resources are found in almost every region of the world, which leads to a degree of energy independence. The choice of energy source has varied. Coal is used for electric power production whereas petroleum is largely employed as a transport fuel, especially since the internal combustion engine came to dominate road transport. Natural gas can be used for heating, and it is also a lower carbon-emitting substitute for coal. The movement to replace fossil fuels with renewables depends upon the ease with which todays' technologies allow for the low cost capture of energy from the wind and sun and its transfer to human markets. As wind and solar farms become common and their energy costs begin to compete with that of fossil fuels, the economic market should start shifting to the new fuel, although perhaps only gradually, since the mechanics for using the new energy source will need to catch up with the available supply of energy.

There are several approaches whereby a society will undergo the replacement of fossil fuels with renewable energy. The market will promote replacement if the costs of using renewables become lower than those of fossil fuels. Government policy can promote conversion from fossil fuel to renewable energy use by taxes or regulations that prohibit the uses of certain energy sources while encouraging the use of others. A carbon tax on fossil fuels would provide incentives in the direction of discouraging fossil fuel use. For example, restrictions that drive up the costs of power from traditional fossil fuel sources or gasoline for transport will encourage other power sources. The other side of this approach is to bring down the costs of preferred energy sources, such as subsidies for renewables, which reduce their costs to power producers, with some of these cost reductions being passed to the consumer to promote demand. There are already federal and possibly state income tax incentives given to consumers for purchasing hybrid and/or electric cars as well as solar panels. There are also rebates for purchasing energy efficient appliances and heating/cooling systems.

In the United States, there has been a concerted political effort, so far unsuccessful, to impose a "carbon tax" on certain fossil fuels.[7] From the economist's view, such an approach would generally be desirable since it would impose a cost on the producer of the external environmental emitter of GHGs. Taxing GHGs will promote measures to reduce their use and therefore their emissions, as well as provide incentives to shift to alternative energy sources, like renewables. Some European countries have successfully imposed carbon taxes.

It should be noted that if the tax is on the GHG directly, rather than on the extraction of the raw material and production of the fossil fuel, a technology that would remove damaging emissions might allow for the continued use of the energy from the fossil fuel. For example, some pollution control techniques allow pollutants to be captured in the smokestack. A technique called Carbon Capture and Storage (CCS) allows fossil-fuel energy to be utilized without the release of the carbon dioxide, since the carbon dioxide is captured at the tail-end of the process, before it reaches the atmosphere. This approach could be viewed as the ultimate clean coal, with essentially no carbon released to the atmosphere, and therefore, no additional carbon taxes.[8]

An alternative to the carbon taxing approach is to try to promote the replacement of fossil fuels with nonemitting renewables. This involves what is called a "carbon trading system."[9] Instead of taxing emissions, these systems require carbon credits or "payments" to allow the release of GHG emissions. The credits are created by activities that capture carbon and/or reduce carbon emissions. These credits can be sold. If an entity is short credits, for example, it can buy them from organizations that are producing credits and have excess credits. These can be sold to firms that require credits. The total amount of credits authorized in the system may be monitored and decrease over time, thus forcing the system to reduce total emissions. A system with some similar features is currently practiced in New England and California.[10]

Techniques drawing from the ideas above have been suggested for forestry. Since trees are an effective way to capture carbon, growing trees can be a tool of reducing atmospheric GHGs. The concept is to manage new forests for carbon capture and make payments according to the carbon captured. One approach would be to allow the forest ownerships to sell forest carbon credits from new forests to be used by entities to meet their required carbon credit needs. Thus a firm would be allowed to release GHGs since it is offsetting those releases with newly created forest carbon credits.[11]

SOME ALTERNATIVES TO FOSSIL FUELS

Addressing global warming by a dramatic reduction in fossil-fuel energy will require the substitution of some other non-GHG-generating energy source. This will involve a shift to other energy sources, both globally and nationally. According to the United States Department of Agriculture (USDA) Energy Information Agency, nearly 90 percent of the world's energy consumption in the late 1960s was derived from one of the three major fossil fuels—coal, oil, and natural gas. That number fell dramatically with the onset of the "oil crises" brought about by the Organization of Petroleum Exporting Countries (OPEC) in the early 1970s, but has largely stabilized near 80 percent thereafter, despite increasing efforts to continue the trend toward renewables.[12] Even now, at the end of the second decade of the twenty-first century, that figure is still very similar as a portion of the whole. Although projections vary, the Energy Information Agency (EIA) forecasts that fossil fuels will still provide about 80 percent of the world's energy in 2040. Thus, although renewable use is increasing rapidly, it is from a very low base. As overall global demand for energy continues to increase, base demand is still being predominantly met by fossil fuels. Of course, the fossil-fuel energy has always been supplemented by other energy sources, including non-GHG-emitting energies such as nuclear, hydropower, and biofuel. However, the potential of these sources as replacements for fossil fuels is undoubtedly limited.

Hydropower

The potential for the expansion of hydropower as a replacement for fossil fuels is probably limited, as a dramatic expansion of hydropower would face formidable obstacles. Although hydropower has great potential in some locations where appropriate water flow conditions exist, such as the Three Gorges Dam in China, these situations are the exception rather than the rule. Although water is almost omnipresent, conditions for hydropower are not. Location is important, not only in relation to water sources but also with regard to locations of power demand. Power transportation costs rise rapidly with distance. Additionally, there appears to be growing public hostility, particularly among environmentalists, against massive projects that often destroy highly valued natural wonders.

Nuclear

Another apparent candidate is nuclear power, which has both the desirable property of zero GHG emissions and can provide a stable power flow. However, a major expansion of nuclear power to replace fossil fuels is unlikely in many countries. There are at least two reasons for this. First, local and global concerns about the possibilities of a major nuclear power catastrophe have created major political opposition to the increased creation of nuclear facilities. This follows in the wake of a number of significant problems associated with nuclear power facilities, including the Three Mile Island incident in the United States (1979), Chernobyl in the former Soviet Union (1986), and the Fukushima incident in Japan (2011). Recurrent disasters in different parts of the world on differing nuclear power systems suggest that nuclear power debacles cannot be assumed to be wholly preventable. Thus, even with new technology, nuclear power will continue to be perceived as constituting a significant threat to global safety. In the United States, waste disposal problems associated with nuclear power have continued to be substantial. The problems are importantly driven by political considerations and include the NIMBY problem: "not in my backyard." Thus, in the United States there has not yet been selected a major nuclear waste depository. Even if one were selected, the logistics of transporting nuclear waste safely from facilities around the country would be formidable, and the need to protect the site against leaking radiation and even possible terrorist attack over the centuries it would take for the nuclear material to deteriorate naturally would be enormous.

Finally, the costs of nuclear energy have become increasingly noncompetitive. Nuclear power costs in the United States are higher than often recognized. Disposal costs, which are not borne by the producers but are transferred to the federal governmental, constitute a growing problem of nuclear power. Recent reports indicate that no new plants are expected to be built in the US as construction and operating costs becoming prohibitive. Due to these costs, as well as safety issues, there is every reason to believe that nuclear power is likely to provide less energy than had been expected in much of the developed world. However, nuclear energy appears likely to still continue to expand in some countries, mostly in the developing world.

What Role for Nuclear Power?

At one time nuclear power was viewed as the energy source that would fuel the world economy. Nuclear power offers the great advantage of minimal emissions of traditional air pollution and zero carbon/GHG emissions. It has the potential to effectively replace fossil fuels with a GHG emission-free energy source. Although not a renewable, nuclear has the advantage of being a stable energy source not dependent on the vagaries of the weather. Of course, a dominant issue with nuclear energy is that of safety and public confidence. Nuclear power is currently in an ambiguous situation, held in disrepute and resisted in much of the developed world, while still seen as an important current and future energy source in many other countries. The divide relates heavily to the developmental stage of the country considering it as a fuel source. Developing countries tend to be looking to expand their nuclear energy both to expand their energy capacity and as a replacement for fossil fuels, while many developed countries are in the process of disengaging from nuclear due to heightened concerns about safety risks. This puts nuclear in the position of being a critical wildcard in the world's energy future. As such, nuclear power is scheduled to play a major role in the future energy plans of both China and India, while being phased out by countries like Germany even as neighbors, like the Czech Republic, are increasing their nuclear power.

The elimination of nuclear power as an option in much of the world, however, has implications for the world's ultimate mix of power sources. First, the reduction of nuclear will place substantially greater demands on renewables to fill the energy gap. Second, a stable and continuous source of power is desired. Nuclear can provide the stability that renewables cannot. The absence of both fossil fuels and nuclear power reduces the ability of a country to provide that stable flow of power. This is because renewable power, dependent upon the wind and sun, is quite variable. For example, if renewable power is at low levels due to weather conditions, the strain on the stable capacity—fossil and nuclear—becomes great and could exceed their capacity, leading to brown-outs. This problem would require excess capacity in the stable power sector. Eventually, technology may allow batteries of sufficient capacity to produce the energy backup, but that capacity appears to be many years in the future.[13]

Note that for Germany much of its recent increased use of renewables to replace nuclear power did not reduce GHG emissions, since no additional GHG emissions were displaced by replacing nuclear. This experience demonstrates the importance of balanced capacity and the limited carbon-reducing ability of renewables when replacing nuclear.

Biogenic

Waste and biogenic material, such as wood, make up about 10 percent of the world's energy production by some estimates, much of it in developing countries and associated with the wood-processing industry.[14] Although energy fuels such as ethanol and methanol produced by plant materials including grains, and wood is being used for power in some applications, as in the wood-processing sector, the potential for biogenics to replace fossil fuels has limited applications. Some European countries have met their GHG targets with the substantial help of wood for energy, which has replaced large volumes of fossil fuels, primarily coal. However, wood itself is a form of carbon, with carbon constituting about 50 percent of its dry weight. So, in the short term, the GHG emission savings is offset by that released by the wood. However, if the tree from which the wood is obtained is replaced, as the tree grows over time, the tree's wood volume is replaced. This net effect is the capture of an equivalent amount of carbon as was released in the burning. Thus this energy was obtained without a net long-term release of carbon, and an associate amount of fossil fuel was not burnt. Of course, a tree will burn in a relatively short time and have a more immediate impact on GHGs, while the newly planted tree will take many years to take out the CO_2 produced by the burnt wood. If the stock of forest wood is expanding, the implication is that growth is exceeding the draw from the forest at that particular point in time. In most forests today, the immediate release of carbon is offset by the sequestration of the many growing trees planted or regenerated earlier.[15]

The intensive use of wood, however, has engendered a debate regarding the extent of tree harvesting for fuelwood and its impacts on forests. Environmentalists express concerns about the extent to which North American wood, which is providing a large portion of the wood being used in Europe as carbon-free energy, is being harvested and redirected to Europe. What are the forest health implications of these levels of harvests? American foresters and forest owners have generally responded by pointing out that forest populations are dynamic. Individual trees, like other biological creators, go through life cycles of birth, life, and death. If harvested trees are replaced and allowed to reach maturity, the long-term health of the forest need not be in jeopardy. Indeed, replacement is required for healthy forests. Supporters also note that despite the increasing draw that Europeans are taking from the US forests, the volume involved is only a small fraction of total US industrial wood harvested, which is mostly used for pulp and paper. North American wildfires have also had negligible impact on aggregate wood stock and harvesting.

The net effect on GHGs is small, since in the US regrowth has systematically exceeded wildfire losses over the past 70 years.[16] Total US forest stock volumes have been expanding continuously since 1950, despite both the effects of wildfires and the rather substantial volumes of wood harvested each year.[17]

In sum, traditional complements to fossil-fuel energy sources will continue to have a role in energy production, but are unlikely to have a *major* role in replacing fossil fuels. Hydropower is constrained by locational considerations, and while important, is unlikely to fill the gap left by the withdrawal of fossil fuels. Although nuclear appears to be almost the perfect replacement for fossil fuels, with a host of desirable traits including GHG free and a stable power capacity, safety issues raise serious red flags. It appears clear that many countries of the world will not move into nuclear; however, overall changes remain to be seen. Finally, biogenic products will likely continue to replace some fossil fuel, but their overall role will be modest. Therefore, the lack of viable alternatives suggests that the long-term solution to the successful implementation of Plan A is the rapid substitution of newer renewable energies, such as increasingly efficient solar and wind, for fossil fuels.

RECENT ACTIONS TO ACHIEVE GLOBAL MITIGATION

Within the context of the UNFCCC was the GHG mitigation effort called the Kyoto Protocol. First agreed to in 1997, it took until 2005 for the protocol to become effective. The protocol called for individual countries to reduce emissions to from 5 percent to 7 percent below 1990 levels by 2012. However, over one hundred developing countries, including China and India, were exempt from country targets. The absence of more broadly allocated targets caused some countries not to participate. Participation was rejected by the US Congress, and the United States withdrew support for the protocol in 2002. The success of the protocol was mixed, with many countries achieving their targets, including the European Union, while others failed. Although not a participant, the United States significantly reduced its emissions level by 2012, but did not reach the 7 percent reduction target that it would have received within Kyoto.[18]

Although some progress was made under the protocol in shifting from fossil fuels to renewable energy, global progress has been quite modest. Many of the reductions could not be said to be "additive," in that they would have occurred anyway. This is particularly true of energy changes following the fall of the Soviet Union and the former Centralized Economies—the old Eastern Bloc countries—where old and inefficient energy facilities were modernized and updated, as in the former East Germany. There were also changes in the United Kingdom as they converted from coal to North Sea natural gas. The United States showed a substantial decrease its emissions largely due to new sources of natural gas from fracking[19] that replaced coal for power plant energy.[20] But despite some successes, global GHG emissions continued to increase rapidly over that period and no changes in the long-term trend are discernible.

PROJECTED MITIGATION EFFORTS

Despite some disappointment with Kyoto, the first major global effort at containing GHGs and global warming, the UN has continued fostering a subsequent accord called the Paris Agreement (2016). Again, the basic approach is one of a centralized, directed mitigation with the intent to prevent or contain the release of GHGs into the atmosphere. Collectively determined individual country targets were not part of the direct strategy this time. The targets were voluntarily determined by the individual countries. A major component of the approach was to offer substantial transfers of income, $100 billion per year, from wealthy countries to developing countries. These transfers were envisaged to finance alternatives to fossil-fuel energy and promote substantial and overall GHG emission reductions. Critics, however, have argued that this transfer was a disguised form of foreign aid.

A coming challenge is found in efforts to stabilize emissions from the developing world, including the two most populated countries, China and India, which are also the world's first and third largest GHG emitters. Both continue to rapidly increase their emissions growth. Between 2005 and 2015, China's emissions have increased by over 3000 million metric tons annually, and India's over 1000 million metric tons annually.[21] By contrast, US emissions levels have declined by over 600 million tons annually over that same period. The rest of the developed world also has experienced significant declines.

The experience of the United States and other developed countries suggests that emissions of carbon and GHGs can, both in principle and practice, be reduced. However, the prospects for stabilizing emissions in the developing world by the 2030 target are dubious. This is especially true for growing countries like China and India, which are intent upon achieving and continuing rapid economic development. Given the preponderance of low-income emerging nations intent on economic development, this suggests that using a mitigation/reduction strategy may be a much greater global challenge than often recognized. Overall, global carbon emissions from fossil fuels have significantly increased since 1970 and have continued their upward trend essentially unobstructed. By 2015, global carbon dioxide emissions rates had increased by about 90 percent over their 1970 levels.[22]

CHALLENGES FOR THE DEVELOPING WORLD

For mitigation to be successful, developing nations need to play a critical role. We know that emerging countries are undoubtedly going to continue to have increasing demands for energy, most of which will be met by fossil fuels, at least in the immediate and near future.

The Paris Agreement[23] provided an increased role for the developing world. Although many of these countries did not have GHG targets in the prior Kyoto Protocol, they were encouraged to select voluntary targets within the Paris Agreement. As noted, a major feature of the Paris Agreement was the financial assistance promised to the developing world to facilitate its transition to renewable energy.

China, the world's leading emitter of carbon dioxide, has experienced an increased rate of emissions in recent years as its economic activity expands from an already high level. In the past it has been a country that has been reluctant to constrain its energy demands. Recently China has agreed to constrain and cap its carbon dioxide emissions and associated fossil fuel use, but only after 2030.[24] Today about 70 percent of China's total energy consumption, and nearly 80 percent of its electricity production, come from coal, the most GHG-emitting of the major fossil fuels. Up to 2030, the agreement also allows China an almost unconstrained increase in use of fossil fuels.[25] Since its achieved 2030 level of GHGs will constitute its future cap, China has an incentive to accelerate its use of fossil fuels right up to that time to get the most production out of its economy before the emission limits kick in. This will maximize its energy options into the future. Some analysts believe China's emissions will peak before 2030; however, according to the Climate Action Tracker (CAT) and based on progress to date, China's target "is not ambitious enough to limit warming below 2 degrees C, that is the limit in the Paris Agreement." The CAT has rated China as "highly insufficient."[26]

Current Chinese plans call for increasing its coal power plant construction well into the next decade. At the same time, however, China has been plagued with massive air pollution problems and has mandated a prohibition on coal for home heating, calling for a shift to natural gas. Unfortunately, an inadequate natural gas supply has created major heating problems.[27] Although this is likely only a temporary problem, it is indicative of the difficulties involved is shifting a massive but still developing economy from one energy source to another. China is also a major nuclear power country, third in the world in nuclear power production, with nineteen new nuclear power plants under construction. It is the fastest expanding nuclear power producer. This will hold China's GHGs somewhat in check. However, China has become an international builder of coal power facilities in parts of Asia and Africa, which implies more GHGs will be coming from these sources.

Following China is India, the world's third leading emitter of GHGs and expected soon to overtake China as the world's most populous nation. It is an emerging behemoth whose demands for energy will surely continue to grow. Recently India's demand for energy—especially fossil-fuel energy—appears to be just beginning to take off. India has pledged to reduce its carbon emission intensity—the GHG emissions per energy unit—and its targets for 2030, according to CAT, appear to be consistent with achieving only modest temperature increases (under 2 degrees Celsius). However, the plans are very general and sketchy, and so accomplishing them remains to be seen. Unlike many countries of the world that are moving away from the use of nuclear power, India plans to expand its nuclear power energy capacity by at least ten new power plants, as well as adding substantial new coal power facilities. Coal is likely to be India's major energy source into the indefinite future.

Obviously, there is much of the developing world that is looking forward to economic development, but development would necessarily involve much higher energy usage. Development without large volumes of fossil-fuel energy does not seem to be a real option for either China or India, at least not in the near or intermediate term.

RENEWABLES: THE SALVATION?

According the more hopeful view, the salvation of the world from warming is renewable energy. And the developing world is involved in utilizing renewable energy. China has undertaken some renewable energy efforts, particularly with wind energy, but overall it is still relying increasingly on natural gas, much of it imported. India's progress with renewables also continues but from a low base, and includes biopower, wind, and solar. Its renewable energy targets are very ambitious and rely heavily on treating hydropower as a renewable in order to achieve its energy goals. Obviously, there is much of the developing world that is looking forward to economic development that would necessarily involve much higher energy usage. The extent to which development can be achieved without substantial increases in fossil fuel use remains to be seen. Serious development without large volumes of fossil-fuel energy does not seem to be a real option for either China or India, at least not in the foreseeable future. Although renewables are expected to provide the energy of the future, and some significant cost-reducing renewable technologies have been developed, the task of converting the energy structure of a single country primarily to renewables is quite daunting. The same task associated with a global transformation is almost unimaginable.

The United States is one of the leading developers and users of renewable energy and has stabilized its GHG emissions. However, US emissions, as well as those of other developed countries that have stabilized their emissions, remain at very high levels. Note that stabilization does not imply that the emissions have ceased. Rather, it implies that the continuance of the flow of GHGs into the atmosphere at a stable but quite high and continuing level; it's just that the level is not increasing but remains relatively constant. Despite the growing use of renewables, which constituted about 10 percent of US energy in 2015, the United States still continues as the second leading emitter globally, of carbon and GHGs. In 2015, America still generated around 80 percent of its power from fossil fuels.[28] Although bioenergy has remained fairly stable at roughly 5 percent of the US total energy, solar and wind have been expanding.[29] But despite a greater emphasis in the potential and adoption of renewables, they have made only a modest dent at best in the total use of fossil fuels in the United States. This is similar in much of the developed world. For mitigation to be effective, the substitution of renewables for fossil fuels must occur at a much more rapid rate than that of the last decade.

The biggest factor in the success of the US stabilization of emissions through the second decade of the twenty-first century has primarily been the shift from coal to natural gas in many applications. However, as noted, natural gas is not a GHG-free fuel. It does have an important advantage over coal, emitting only one-half of the carbon dioxide per unit of energy of coal, thus allowing for large declines in GHG emissions. Thus, natural gas is commonly viewed as a bridging energy source that will allow emissions to be gradually stabilized as efficient technologies of renewables—especially wind and solar—are developed.

A SUCCESS?

An example of a major country that has been successful in moving from fossil fuels to renewables is Germany. Germany produces over a third of its energy from renewables, including wind, solar, wood biomass, and other biomaterials. Germany has invested heavily in offshore wind and expanded wind power to produce an expected one-third of its energy in the future. It is estimated that Germany will have spent over $222 billion on renewable energy subsidies between 2000 and 2019. However, not all of this power replaces fossil fuels, since some of it replaces nuclear, which Germany is also moving to eliminate. Given its variable demands on energy, at times Germany has produced so much renewable energy that it has a problem using the power produced from its more traditional power facilities. At the same time, some facilities, such as coal plants, cannot dial back production quickly enough to prevent excess production from occurring. At those times Germany has subsidized its consumers (literally paid them) to use its energy just to keep the traditional facilities operating.

Despite its stellar performance converting from fossil fuels to alternatives, Germany actually increased its GHG emissions in recent years (2016). The loss of non-GHG-emitting nuclear is the cause. Since some of the renewable energy sector, like wind and solar, are subject to large weather-driven fluctuations, the national system needs to work off the stable energy component that provides a buffer to provide power, even when the energy from renewables is small or absent due to the weather. In Germany, this buffer is largely coal driven, given the absence of nuclear. These peak power and transitional power issues complicate the management of the various large-scale energy-producing systems. In the future, the development of large-scale batteries to capture excess energy may be required for these systems to function smoothly. So, despite major progress and successes, Germany has reduced its 2020 climate target to decrease its GHG emissions from 40 percent to 32 percent. These targets of reduced emissions have been moved further into the 2020s.[30]

A basic tenet of the mitigation approach is that renewables will replace fossil-fuel energy. Additionally, we see a strong movement by many developed countries to eliminate nuclear power. Above I note some of the difficulties being faced by Germany. This type of problem is not limited to any one country. As suggested by Raimi and Krupnick,[31] a widespread problem is that of putting in place adequate renewable power facilities to replace both fossil-fuel energy and nuclear. An example of this issue is found in New York State. The governor has announced a $6 billion program to build 2,400 megawatts of offshore wind capacity by 2030. Simultaneously, the governor is opposing the expansion of nuclear power facilities for the state and planning to close the Indian Point Energy Center by 2021.[32] It is estimated that the wind plants will produce about 60 percent of the power now produced by the nuclear plant. What will fill the hole in the stable portion of the state's power facilities? In New York State, they are looking to natural gas.[33]

CARBON DIOXIDE CONTINUES TO RISE

Although wind and solar energy are gradually replacing some fossil fuels in certain applications in parts of the world, carbon dioxide and GHGs in the atmosphere continue to rise. Atmospheric carbon dioxide is now over 410 p/m, a record high, even as fossil-fuel energy continues to account for over 80 percent of the world's constantly growing energy demands. Although data suggests that a slowdown in carbon buildup may have occurred in the second decade of the twenty-first century, most analysts attribute much of this to the global economic recession in those years, beginning in 2007/2008. By the best accounts, the global temperature continues its substantial rise, despite evidence of a short-term hiatus from 1998 until 2016.[34]

LIMITS OF MITIGATION

Today, the world is focusing on a mitigating approach to deal with climate change, which had gained the favor of Al Gore and others in the early 1990s. These policy leaders became alarmed by environmental changes and indications of climatic instability. Their proposed solution has been to utilize the UN to focus the international community on stopping the emissions of GHGs and shifting our energy systems away from fossil fuel and toward cleaner energies.

But how feasible is this approach? Is this "a bridge too far," as with the failed attempt by the Allies in WW II to capture the bridge at Arnhem, a task that was beyond the capability of the armies then? The data indicates no serious moderation in the overall growth of fossil fuel use or GHG emissions over the past thirty years, the period since the FCCC has become active. Projections discussed below suggest that current mitigation efforts are inadequate to achieve the targets. Although some of the developed countries have reduced or stabilized their emissions, the global economy continues its growth and countries like China and India, with well over two billion people combined, are among the most rapidly expanding economies. Fossil fuels continue to be the driver of economic growth, providing over 80 percent of an ever-expanding supply of the earth's human energy. While the United States has stabilized emissions, it is still the world's second leading emitter of carbon and GHGs. Although in the long run renewables appear to have the potential to largely replace fossil-fuel energy, the process is on track to take many decades if not centuries. If warming is the threat that it has been presented to be, the current and anticipated rates of mitigation are woefully inadequate. Even if human GHGs were the sole source of rising atmospheric concentrations and global warming, how likely is it that the global increases can be stabilized, to say nothing of reduced, in the foreseeable future?

There is rising concern about the ability of negative emissions to undo much of the damage that GHGs have generated. The long-term issue of addressing GHGs requires not only stabilizing emissions, but also negative emissions, whereby the earlier buildup of GHGs is reversed and then largely eliminated. This result comes out of an investigation of over one hundred scenarios designed to keep temperature rise below a 2 degree Celsius increase from preindustrial levels. Nearly all of these assume negative emissions technology. However, most analyses find that such an approach is not feasible at the level required to have the desired global impact. A recent study by the European Academies of Science Advisory Committee from January 2018 found that a system large enough to accomplish this end would probably be unsustainable. The study concluded "these technologies offer only limited realistic potential to remove carbon from the atmosphere and not at the scale envisaged in (many) climate scenarios."

The implications of the difficulties discussed above are summarized by Raimi and Krupnick in a "Resources for the Future" blog.[35] They note that "the energy system is enormous and changes slowly" and that "even if (wind and solar) grow rapidly, the sheer scale of the energy system means the most rapid transition would (still) take many decades."

ADDRESSING CLIMATE WARMING REVISITED

As argued elsewhere,[36] a common view is that even if the principal source of warming is GHGs and if the targets of the Paris Agreement are in fact met, the effect on global temperature will be "tiny." It has been estimated that even with full compliance with the Paris Agreement, reduction of the global temperature by the year 2100 will be far too small, about 0.2 degrees Celsius, to meet the temperature target of holding warming to less than 2 degrees Celsius.[37] Other estimates are that the fulfillment of Paris pledges will only provide a third of what is needed to hold the temperature below 2 degrees Celsius by 2100. [38] In fact, most countries are falling short of even their carbon mitigation targets. This view is consistent with the data presented above, showing little global progress in reducing GHG emissions while noting that the preconditions exist for large increases in emissions well into the next several decades. Supporting this conclusion is the reality that economic growth in the populated developing world is contributing to continued high use of fossil fuels for energy and accordingly minimal success at reducing GHG emissions. Thus, progress in reducing GHG levels and emissions is proving to be painfully slow and there is little realistic possibility for this to change over the next several decades.

Finally, any warming from natural climate variation can be expected to continue unabated by Plan A, since most evidence suggests natural warming is mostly unrelated to atmospheric GHG emissions. Given that mitigation is clearly inadequate to controlling GHGs and that any natural warming is likely to be due to non-GHG factors, it appears that at some point in this coming century much more emphasis will need to be directed to a different approach such as adaptation. However, adaptation, defined broadly, includes a host of approaches that anticipate global warming and aims to offset the damage associated with it. Part of this process also involves geoengineering approaches geared to modifying the atmosphere to neutralize any increased warming capacity due to human activities, which entails humans intervening in such a fashion as to offset any activities that may alter the atmosphere and promote warming. This includes effects that would occur if the atmosphere were left to simply accumulate increases in human-emitted GHGs, as well as natural variations.

MITIGATION: SOME MORE PROBLEMS

Of the two basic approaches to addressing climate warming, it seems likely that the mitigation benefits Plan A are likely to deliver too little, too late. At some level, however, this approach seems to be appropriate. If we believe we understand the nature of the phenomenon that is driving the warming, and that humans have been major contributors to it, we should have the capacity to control it. In the case of the current warming, the conventional wisdom puts a great deal of confidence in the hypothesis that warming is caused by human emissions of carbon and other GHGs. These emissions come overwhelmingly from the fossil fuel system. The apparently obvious resolution is to cease emitting GHGs, an objective that seems to be best approached by a restructuring of the global energy system. However, as is now widely recognized, this is easier said than done.

The organizational approach would be a centralized, top-down approach, with the UN taking a critical role. An agreement would be created, and the UNFCCC would endeavor to rally the world's nations in a common effort to address the warming problem, resulting in a global initiative to try to reorient the earth's energy system away from GHG-emitting fossil-fuel energy and toward non-GHG-emitting energies such as renewable fuels. However, the success of this approach has been limited thus far: the global rate of growth in emissions has increased since the beginning of the twenty-first century, even as developed countries have moved to stabilize their emissions. The force driving the buildup has been developing countries' use of fossil fuels in their quest to achieve economic growth goals. This global growth in emissions has occurred in spite of efforts by the UN and other global entities to increase their development and implementation of various technologies to support renewable energy. It is not that renewable energy is not being harnessed and rapidly utilized in some countries. Rather, it is simply that the total demand for energy, largely in developing countries, has overwhelmed the aggregate energy supply, including the very modest increases in renewable energy. This rapid increase in demand is likely to continue for several decades as developing countries gradually experience economic growth. Most of the demand increases will almost surely be met by fossil fuels, since the infrastructure for their use is already in place.

Efforts using mitigation techniques are not reducing amounts of GHGs fast enough. The Climate Action Tracker (CAT), which monitors emissions and estimates temperatures consistent with these GHG emission levels, has stated future expected emission levels and compared them with the temperature goals of the Paris Agreement. It notes that "a substantial gap remains between the expected levels of emissions in 2025 and 2030 and the lower temperate goals in the Paris agreement. While the temperature goal is to limit warming to 1.5 to 2.0 C, unconstrained practices would result in predicted temperatures that are 4.1 to 4.8 C by 2100. If current policies were continued this would bring these down to 3.1 to 3.7 C, or roughly a one degree reduction to the year 2100."[39] Plan A may be inadequate to addressing the threat that humanity faces.

In the "Presidential Climate Action Project" section on "Emissions Reduction Needed to Stabilize Climate" by Susan Joy Hassol, question number four asks specifically: "How much and how fast do industrialized and developing countries need to reduce emissions in order to meet this temperature target?" The answer the short report provides follows:

> In order to stabilize CO_2 concentrations at about 450 ppm by 2050, global emissions would have to decline by about 60% by 2050. Industrialized countries' greenhouse gas emissions would have to decline by about 80% by 2050. One way of doing this, proposed recently in a study

by scientists at Duke University's Nicholas School, would be for the G8 countries to decrease emissions by an average of 2% per year starting in 2011, using 2010 as a baseline, for 40 years, resulting in an 80% reduction by 2050. In addition, the five largest developing countries (China, India, Brazil, South Africa, and Mexico) would begin a similar program ten years later, reducing their emissions by 2% a year starting in 2021, using 2020 as a baseline, and the rest of the world would have to stabilize emissions between 2030 and 2050. Around midcentury, most other nations would also have to begin reducing emissions and/or the initial G8+5 would have to make greater reductions, in order to maintain the CO_2 concentration at or below 450 ppm. In addition, we would have to reduce emissions of the non-CO_2 greenhouse gases (methane, nitrous oxide, soot, halocarbons, etc.) sufficiently to keep the temperature below the threshold discussed above. If we did not reduce those gases as well, their warming effect would take us above the target temperature discussed above. An analysis by Meinshausen shows that the achievement of long-term stabilization at 400 or 450 ppm CO_2e[40] will imply a temporary peaking of CO_2e concentrations at higher levels, for example, peaking at 475 ppm and then stabilizing at 400 ppm roughly a century later. Following such a path would result in about a 75% chance of keeping the temperature rise below the 2°C/3.5°F target.[41]

Note that the task project response mimics a "Mission Impossible" type response. Although perhaps technically feasible, creating the political will in all of the different countries, with their many factions and broad desires to achieve economic development and growth, and harnessing individual energy to move in such a direction is extremely problematic. Today, the CO_2e is about 410 p/m, within the critical range outlined in the articulation of the challenge of the report. That means the global atmospheric GHG levels are currently in the range consistent for meeting the less than 2 degree Celsius goal. However, the targeted date is 2030, and the world is still on GHG ascendency. Thus, not only is stability required immediately, but reversal is required to meet the temperature objectives. However, the world is still using about 80 percent fossil fuels to meet its energy needs, and large numbers of conventional coal power facilities are being undertaken and planned in China, India, and elsewhere in the developing world. According to the charge above, the developed world should have been reducing its emissions by 2 percent annually beginning in 2011, while the large developing countries should be getting prepared to begin 2 percent declines. Clearly, neither of these events have happened nor do they appear to be beginning to happen. The prospects of seriously addressing global warming with the current plan, essentially a form of Plan A, are almost nonexistent. Hence, more is needed and there is an opportunity, indeed a necessity, for Plan B, both as an adaptation to climate and the damage likely to be forthcoming, and as a supplementary approach to contain warming by offsetting at least some of the changes in the atmosphere that are generating the warming.

RESEARCH EMPHASIS

In my view, mitigation could more constructively be shared with adaptation. Fortunately, some local activities are now being oriented toward on-the-ground adaptation (see Chapter 4), but adaptation is still largely being disregarded. It should be noted that the emphasis on additional approaches other than mitigation in addressing climate change have also been mostly ignored by the UN and FCCC.

Other possibilities that need to be utilized, as well as general adaptation, include geoengineering. As discussed in Chapter 5, geoengineering appears to offer a substantial potential for addressing much or perhaps all of the warming problem. The general idea embodied in many geoengineering approaches is not to prevent the release of GHGs, but rather to neutralize their warming effect once they have been released. The basic idea is to change the reflectivity of the earth system, making it less likely to absorb the heat of solar rays. Some of these approaches are quite controversial. Current funding levels for geoengineering research has been estimated at levels as low as $10 million annually.[42] If the damage levels likely to be associated with temperature increases that move above the 2 degree Celsius ceiling are anywhere as draconian as often portrayed, rational behavior would certainly dictate that some research should be undertaken in the alternative geo-pathways that might provide useful technology. As this chapter suggests, our human ability to avoid temperature increases well above those targeted is not guaranteed by current mitigation efforts, and indeed may be unlikely. Additionally, conducting studies to try to assess the true risks associated with a particular applied mitigation approach would certainly be sensible. As suggested by David Archer,[43] the global carbon cycle is not well understood, and this implies significant uncertainty and inherent risk in society's current mitigation approach. As with any new technology, unintended consequences are certainly not unique only to a geoengineering approach.

At costs estimated at something in the neighborhood of $10 billion per year, technical geoengineering approaches appear to offer the potential to stabilize,[44] or assist in the stabilization of, the earth's temperatures by modifying the atmosphere so as to offset the additional warming that the increased GHGs generate.[45] This cost is well under the $100 billion per year built into the Paris Agreement, which only covers the transfers to developing countries and does not consider the additional costs of early conversion from obsolete fossil-fuel energy systems to renewable systems. As noted elsewhere (see Chapter 5), although the geoengineering topic has been covered in the IPCC's assessment and other reports, it has been given only very superficial and perfunctory coverage.

General adaptation and adaptive management is also not well-covered by the IPCC and its Climate Assessment Report. I note that there are relatively few studies on efforts to deal with climate warming that pay any serious attention to adaptation responses. Even in the case of jurisdictions like California, which is very sensitive to climate issues, the lion's share of the emphasis and resources are directed toward efforts to mitigate carbon emissions. At one level, this is curious for California, and indeed most local jurisdictions, since even if these entities totally eliminated their individual GHG emissions, the effect on global warming would be negligible, since they are such a small fraction of the global total. By contrast, redirecting a portion of those resources to adaptation could finance a renewed infrastructure more capable of anticipating and addressing climate warming and directly more benefit the citizens of the state.

Let me point to a personal experience that suggests a bias in the decision process related to adaptation approaches. At Resources for the Future, an independent think tank where I was employed for over thirty years, we had a resource and climate program in the late 1980s that undertook a project to examine adaptation potential in areas of the US heartland that had been devastated by the dust bowl of the 1930s.[46] The effort was funded by the Department of Energy. The concept was to examine the Midwest to estimate what would be the consequences of another dust bowl climate period. The states investigated were Missouri, Iowa, Nebraska, and Kansas. The focus was largely on agriculture, including forests and grasslands. We observed that when we presented the project as related to climate and examining adaptation, we were counseled to downplay both the connection to climate and the adaptation aspect. At that time, adaptation was treated as "politically incorrect." Upon completion of a very successful project from a research-finding perspective, we submitted an application for a follow-on effort with strong adaptation aspects. However, we apparently were unsuccessful in camouflaging our proposed adaptation project and received indications from the funding source that adaptation was not a priority area and subsequently the funding was not forthcoming.

Although the priority given to adaptation appears to have risen a bit over the intervening three decades, the generally low priority given adaptation appears to continue to this day. This is reflected in the IPCC reports, US research funding efforts and levels, and the mix of climate control activities being undertaken across the United States and much of the world.

SUMMARY AND CONCLUSIONS

Recently a series of scientific papers have stated that we have only a 5 percent chance of limiting warming to 2 degrees Celsius, and only a one percent chance of limiting it to less than 1.5 degrees by 2100.[47] This is assuming that society attempts to move forward with the Paris Agreement, the active variant of Plan A. This mitigation approach assumes that the basic problem of global warming is due almost entirely to carbon and other GHGs released by human actions and the emissions are therefore amenable to human alleviation. However, mitigation is a less viable option when dealing with climate change that is, in part, driven by natural forces. There is substantial evidence for the existence of a natural variable driver, although the essence of that is both controversial and not well understood. Even if a mitigation approach were to address the GHG warming that is driven only by human activities, it does not appear likely to be able to stop the warming.

A second limitation is found in society's inability to adequately substitute renewable energy for fossil fuels. Forecasts of the damage that will occur if the mitigation approach is not successful are very severe. The continuing high and increasing use of fossil fuels appears unlikely to abate soon enough to avert continuing significant increases in the atmospheric level of GHGs and the associated temperature increases. Hence, the mitigation strategy is highly unlikely to abate GHG emissions and thus to contain warming soon enough to avoid the vast amount of damage that is projected. If the potential for natural forces to promote climate change is factored into the assessed overall risk associated with a mitigation approach, the degree of risk increases substantially. So, the likelihood that humans can make the transition to a largely renewable GHG-free energy system solely through the mitigation approach is highly problematic. Only recently has it become clear that mitigation is very unlikely to restrain temperatures to within acceptable levels of between 1.5 and 2 degrees Celsius. If this prognosis is correct and the consequences of failure are as dire as they are portrayed, it is increasingly clear that another approach is needed. That approach is an adaptation Plan B, including the use of geoengineering.

CHAPTER 4

PLAN B:
THE ADAPTATION SOLUTION

I n a 2018 hearing, Federal Emergency Management Agency Administrator Brock Long said his agency is planning around rising seas, and that proposed budget cuts to climate change programs won't affect that. "Obviously FEMA can't stop sea-level rise. That would be the equivalent of us saying we're going to stop plate tectonics as well and halt all earthquakes," Long said. Rather than resist proposals to cut funding for climate programs, he asked instead for lawmakers to focus on what he says the agency needs: more funding for pre-disaster grants, more staff, and more state-by-state flexibility. In short, he called for enhancing adaptive capacity.[1]

Global warming is a worldwide issue. Not only are all the world's countries threatened by the warming caused by GHGs, but the source of these damaging GHGs comes essentially from every country in the world. To address GHGs and climate change, an international environmental treaty to commit to control such emissions, called the UN Framework Convention on Climate Change (UNFCCC), was signed in 1992 by more than 195 countries. The United States is a signatory to this treaty.

An early step within the FCCC process was the Kyoto Protocol, which commits countries to specific targets for GHG emission reductions. Despite some disappointments with the results of Kyoto, the UN then fostered a subsequent agreement called the Paris Climate Agreement,[2] signed in 2015. The basic approach of that agreement is a centralized, directed mitigation with the intent to prevent or contain the release of GHGs into the atmosphere. Individual targeted reductions are selected voluntarily by each participating country and are not part of a collectively determined per country and global target, as was the case with Kyoto. The approach also offers substantial transfers of income, $100 billion per year, from wealthy countries to developing countries to finance climate controlling activities. The United States has not signed on to either of these agreements.

The coming challenge is found in attempting first to stabilize GHG emissions, and then reduce the GHG emissions levels, particularly from the developing world, where they are still increasing. This challenge is formidable, as we have seen, for China and India. Between 2005 and 2015, China's emissions have increased by over 3000 million metric tons per year and India's is over 1000 million metric tons. By contrast, US emission levels have declined by over 600 million tons over that same period, while most of the rest of the developed world have achieved stability, with some countries achieving modest emission declines.[3] The experience of developed countries suggests that emissions of carbon and GHGs can, in principle and practice, be reduced. However, the prospects for emission stabilization by the developing world, and especially emerging economies like China and India, which are intent upon achieving and continuing rapid economic development, are much more problematic. Global carbon emissions from fossil fuels have significantly increased since 1970 and continue their upward trend essentially unobstructed. By 2015, CO_2 emissions had increased by about 90 percent over their 1970 levels.[4]

Even the stabilization of emission levels where their growth is at zero would still involve the continuance of massive and ongoing net flows of GHGs into the atmosphere. The continued global buildup from current levels, which has driven the atmospheric GHG level to a record 410 parts per million,[5] will continue to drive up the heat-capturing capacity of the atmosphere at least into the near future.[6]

PLAN B: ADAPTATION

Given that the principal source of warming is human-generated GHGs, and even if the targets of the Paris Agreement are met, the prevailing view is that the effect on temperature will still be vanishingly small. In fact, most countries are likely to fall short of their agreed carbon-reducing targets. This situation is made worse to the extent that warming from natural variation will continue unabated. Hence, at some point in the coming few decades, more emphasis will need to be placed on adaptation. Damage will surely occur and will need to be overseen.

The likely damages are many, varied, and potentially large. They include loss of habitat, decline of glaciers and associated loss of certain fresh water sources, acidification of the oceans, erosion of coastlines, and perhaps an increase in extreme weather events, as with the hurricanes Harvey, Irma, and more recently Florence. The three types of incidents that I expect are likely to be most destructive relate to the sea level rising, damage to agriculture and agricultural productivity, and damage to forests and biodiversity. In addition, I will touch on some other likely destructive events.

The principal argument of this book is that adaptation should become a major tool to address climate warming. Plan A is simply not adequate to prevent continued warming and protect humans from its ravages because it does not deal with damages that are already inevitable. I argue that Plan B's emphasis on adaptation can provide this. Adaptation or adaptive management in the climate arena refers to a focus on the management of the damage associated with warming, rather than trying to stop the warming itself. The argument is simple. Warming is difficult to stop. We see that even all the efforts directed so far to mitigation of human-generated warming are, by some estimates, expected to reduce temperatures in the year 2100 by only 0.2 degrees Celsius.[7] Although this estimate may be extreme, it is widely agreed that the most likely improvements from current mitigation programs are not nearly enough to prevent the onset of large-scale damage predicted by 2100, including sea-level rises of up to several feet. In addition to GHG emission increases, there are questions of reversing past accumulations of GHGs and neutralizing the effects of natural climate variation. Thus, stopping the increase in emissions of GHGs is only the first step to a broad unwinding of the past buildup of atmospheric GHGs.

Plan B is an acknowledgment that the first response of the global community—efforts at reducing or eliminating human-generated GHGs—is not doing an adequate job of restraining warming by itself, which means that additional measures need to be undertaken. Plan B would reallocate some of those resources and efforts toward adaptation, so as to reduce the total impact by working at the damage-management level as well as at the prevention level.

Adaptation assumes society will respond to perceived threats. So, too, does mitigation that involves substantial investments in reducing GHGs, which are going to generate damage at some unspecified future time. However, we see evidence of anticipatory behavior every day, not only with respect to global warming and climate change, but also in activities related to national defense, crime control, and immunizations. There is no reason not to apply the same principles to climate change. The following sections identify likely damage areas and suggest adaptation responses to anticipated climate-change-generated damages.

ADDRESSING MAJOR DAMAGES

An important element to addressing future climate change damage is the development of new technologies. Major technological strides have been made to make America's critical infrastructure more resilient to natural disasters. New types of concrete have been developed, as well as space-age sensors to detect defects and weaknesses ready for implantation in infrastructure, new plastics, new designs, and new management methods. The collective effect of these leaps is to make infrastructure more resilient to hurricanes, floods, earthquakes, fires, tornadoes, terrorist attacks, and other threats.

And yet instead of adopting these new technologies, governments often rebuild infrastructure the way it was installed decades ago, virtually assuring similar damage in the next disaster. The adoption of new, proven technologies to enhance the resilience of America's critical infrastructure has been painfully slow. To address this, new methods of delivery are required, ensuring that revolutionary technologies to enhance infrastructure resistance are actually adopted by state and local governments.

R. Richard Geddes of the American Enterprise Institute has argued for a public-private partnership (PPP).[8] The core of a PPP is a contract between a public-sector partner that must rebuild its infrastructure, and a private partner (usually a group of firms) that provides the rebuilding as well as the operational and infrastructure maintenance services. He notes that the PPP contract combines, or "bundles," together tasks required to rebuild, while operating and maintaining existing infrastructure. For example, the $3.9 billion contract for the Tappan Zee Bridge replacement north of New York City was a PPP combining the design and construction of the new bridge.

A PPP contract, according to Geddes, increases the adoption of cutting-edge, durability-enhancing technologies in at least three ways:

First, PPPs use the global knowledge and best-practices expertise of private companies to rebuild infrastructure, typically combining in-depth experience with innovative technologies.

Second, PPPs can ensure flexible infrastructure because the group of private firms that design and rebuild the facility will, in the long run, also operate and maintain it. Thus, the private partner has strong incentives to incorporate the best materials, sensors, and new technologies.

Third, the structure of the PPP contract itself creates incentives to adopt new technologies. Such an agreement can simply require that the private partner use certain technologies known to survive a natural disaster or—even better—it can lay out how the infrastructure will perform in "stressed" situations. This allows the private partner to decide which technologies will best meet the performance standards.

According to Geddes, many other countries rely on PPPs. Japan, for example, issues hundreds of contracts with private firms to operate and maintain its roads, bridges, and tunnels. Florida is a leader in PPP use in the United States and is in an excellent position to leverage this method for its current recovery from seasonal hurricanes.

The United States is judged to be decades behind many other developed countries in the use of PPPs but can learn from international experience and the standard practices that have emerged. The incorporation of PPP-type systems will provide for additional damage reduction for future climate events by providing new modern infrastructure and preventing future degradation of current infrastructure systems.

RESILIENCE

There is a tendency for disaster regulations such as the national building standards to require that new structures be designed to at least meet events comparable to those encountered in the most recent disaster. However, the future may be even more severe. In concept, the standards should be designed to meet likely future disaster events. These are not known with any certainty and judgments on the severity of these may vary, even among experts. Given the uncertainties, many local governments and federal entities are advocating precluding rebuilding in hazard zones. But climate change suggests hazard zones may change over time and flood maps are becoming increasingly dated. Note that subsidized hazard insurance, such as Federal Flood Insurance, will promote locations in high-risk areas, though abandoning some of these properties could involve substantial financial losses to owners and they may resist this measure.

In general, experience suggests that people are often willing to incur costs to avoid risks even when the risks are not well-known.

SEA-LEVEL RISE AND FLOOD RISK

In testimony before a Congressional House Appropriations Committee hearing in 2018, FEMA Administrator Brock Long said that his agency is making plans to address rising seas. He asked lawmakers to focus on what he anticipated the agency needs, which was more funding for pre-disaster grants, more staff, and more state-by-state flexibility.

More broadly, there is little question that a rise in sea levels could cause major damage to human life and coastal development as we know it. Actuaries, those charged with estimating insurance risks, find rising seas to be a greater source of damage risk than heat alone. According to the "Actuaries Climate Index," the climate change deviations from a thirty-year average are likely to be greatest in the latter part of the second decade of the twenty-first century.[9] During this period, the East Coast of the United States has indeed faced increased flooding incidents.

A flooding problem could involve the inherent instability of some of the major land-based ice sheets. Scientists have expressed concerns in regarding the West Antarctic Ice Sheet and the roughly three meter sea-level rise that could accompany its collapse.[10] Even without such a collapse, a study published in *Nature*[11] estimates that sea levels could rise at an accelerating rate due to the Antarctica ice sheet melting three times as fast as it did twenty-five years ago.

Sea level has risen and fallen many times over the earth's life, including a slow but continuing rise during this current interglacial period beginning over twelve thousand years ago (Figure 2.3). One of the early explanations of how humans became established in the Western Hemisphere is a low sea level that allowed peoples residing in Siberia to cross over the Bering Strait to Alaska on a land bridge.

The land bridge was the result of reduced sea level at the end of the last ice age, due to the fact that large amounts of water were tied up in glaciers at that early period. Although there is increasing evidence that land bridge migration to the Americas may not have brought the first humans to inhabit parts of the Western Hemisphere, the Bering crossings certainly appear to have been a feasible and utilized migration route. The concept of a gradually rising sea level as the interglacial period progressed is largely accepted by scientists.

In recent periods, sea level has been rising about eight inches per century. More recently however, the rate of rise has increased, although not yet dramatically, to closer to one foot per century.[12] Obviously, these slow increases are difficult to detect without precise scientific instrumentation, and should they continue at these rates, would provide only modest amounts of damage in the short term, mostly associated with storm-related sea surges.

There are a host of projections of sea-level rise,[13] some of which have anticipated an increase of level of up to a two-foot increase by 2050, and others as much as ten feet by 2100. In a recent hearing of the House Science, Space, and Technology Committee of Congress, Philip Duffy, president of the Woods Hole Research Center in Massachusetts and former senior advisor to the US Global Change Research Program, is reported to have stated that "if we let the planet warm 2 or 3 degrees, we will have tens of meters of sea-level rise."[14] The source of the increases would be the substantial rise in the melt rate of land borne glaciers, which is not offset by increased snows that replenish the glaciers, and the flow of these waters to the sea. Additionally, thermal expansion of ocean and sea water could contribute to the rise. If sea-level rise is accompanied by more intensive storms and storm surges, the increasing problems are obvious. However, other scenarios are possible.[15] Global warming is known to be associated with increased precipitation. High levels of precipitation in polar regions could result in higher snowfall levels and could help replenish the loss of glacial ice, thereby offsetting some of the accelerating rise in the sea level.

Projections are regularly made showing increases in threats to coastal areas associated with some amount of sea-level rise as well as with coastal rise or subsidence, where appropriate. At times, projections are made of the estimated financial costs associated with such damage to the infrastructure and the economy. Although the amount of recent sea-level rise is controversial, many projections have anticipated substantial sea-level rises by the mid-twenty-first century. A recent study estimates that thousands of homes are at risk due to sea-level rise along the California coast. Of course, this depends in part upon how successful climate change mitigation turns out to be, and how successful adaptation efforts may be.[16]

Although forecasts of future sea-level rise are ominous, the debate over the extent of recent sea-level rise is still contentious. Sea level is extremely difficult to measure, requiring accuracy down to the millimeter. Coastal tide gauge measurements can be taken along with measurements through sediments and via satellite altimeters. Sea level varies by location, depending upon prevailing winds, tectonics, and gravitational variations, which result in uneven sea-level rises due to subtle differences in gravity.[17] For example, sea-level increases are estimated to be 52 percent greater in parts of Florida due to gravitational variations.[18] Additionally, the earth's surface can vary in the coastal zones. Erosion and subsidence can draw the earth's surface downward while forces within the earth push it upward in some places. Hence, different measurement techniques in different locations may provide different estimates of sea-surface rise, based in part on land surface movements.

Despite such uncertainties in future sea-level changes, a more promising approach than attempts to prevent sea-level rise could be that of applying adaptive management to reduce climate-change-generated damage. Such an approach allows for a customized response depending upon the specific circumstances of the situation. It allows for a more aggressive approach where the damages are likely to be significant, and a more cautionary approach where they are likely to be self-contained. Where an aggressive approach is called for, it might include the building of dikes and other water barriers, as well as improving drainage. A more thoughtful futuristic planning of replacement infrastructure should be developed to anticipate and avoid an area being overrun by the sea.

SOME ADAPTIVE EXPERIENCE RESPONDING TO SEA-LEVEL RISE

A very large portion of the human population is located by the sea. If attempts at preventing global warming are not dramatically successful, sea-level rise can be expected to accelerate. Most projections forecast difficulties with sea-level rise even if humankind does meet the climate mitigation targets currently in place. What types of adaptations might be promising in addressing this problem?

A number of coastal areas are beginning to undertake adaptive measures to deal with current and anticipated sea-level problems. California, for example, is beginning to respond to concerns regarding sea-level rise and coastal issues. Being a seaside state, it has needed to consistently address challenges associated with coastal infrastructure and safety. Recently, however, it has begun to consciously include anticipated sea-level rise in its plans and construction projects for coastal infrastructure and protection.

California has been at the forefront of states primarily addressing climate change through efforts to mitigate carbon and GHG emissions.[19] However, it is beginning to face the prospects of an increasing amount of climate-related damage, primarily in its coastal areas. A recent US Geological Society study projects that California coastal cliffs could recede by 130 feet by 2100.[20] California policy makers are now expecting, at the high end, a ten-foot sea-level rise.[21]

It is estimated that for every foot of global sea-level rise caused by melting ice sheets, California will experience a 1.25 foot sea-level rise due to the earth's rotation and the gravitational pull on the waters.[22]

Larger water rises, such as could occur in California, require extreme adaptations including, perhaps, the relocation of significant populations. In response to these challenges, new bridges near the coastal areas of the state are today being built a bit higher and longer in anticipation of future sea-level rises. The state has produced an updated policy that provides guidance to state agencies for incorporating sea-level change considerations into projections relating to planning, construction permits, and investment decisions.

The anticipation of the continuing rise of sea levels allows and indeed requires continuing adjustments. Infrastructure update and replacement are common, even in the absence of anticipated sea-level rise. Old infrastructure needs to be replaced periodically. Under dynamic conditions of rising sea level, rather than a simple replacement of aging infrastructure, updating could be undertaken on a shorter periodic basis and be of a somewhat different type. New and replacement infrastructure can be installed at more advantageous locations to respond to the more dangerous sea-level rises and adjusted to newly revised expectations. Resilient infrastructure must be developed to replace traditional infrastructure types.

Thus, although the costs of these responses and adjustments will no doubt be significant, some of the costs could be essentially drawn from what would be the already required allocation of maintenance funds. More generally, the shortened useful lives of projects made obsolete due to sea-level rise would allow the reallocation of major maintenance funds to newly occurring developed projects that become necessary to address previously unanticipated but now necessary sea-level-rise-related projects. This reallocation of funds could reduce the overall burden of the adjustment.

The following activities and projects, which explicitly include consideration of sea-level rise, are either underway or under consideration in California:

- building of seawalls and barriers;
- dredging and the continuing of sand replenishment;
- stabilization walls along highways;

- relocation and modification of structures, either currently or when replacement or major renovations are undertaken, located in the coastal zone and adjacent areas with particular attention given to power plants and nuclear facilities;
- infrastructure relocation and modification of bridges and roads, either current or when replacement or major renovations are undertaken;
- installation of drains, rainfall interception networks, wall stabilization structures;
- development technologies for subsurface observation;
- relocation of populations at risk to higher elevations.

The use of all these measures, including dredging to shore up the oceanfront, can be practiced on a continuing basis.

More broadly, however, the general approach also needs to include policies to discourage development in the high-risk areas along the coastal zone. Current policies should be oriented to providing disincentives for location in the risky coastal zone areas and designed to promote new locations and relocations of development in more risk-free areas.

It should be noted that West Coast issues are somewhat different from those of other coasts, since the tectonics are tending to uplift the coastal areas, creating cliffs and bluffs, which are undermined by the sea below. The nature of a managing solution is a continuing process that adapts to the on-going tectonic coastal land movement and continuous sea-level rise.

Rising Sea Level in Del Mar: Del Mar California, a small beach community, is struggling to choose from programs that offer protection from an anticipated sea-level rise. To address the issue, the city of Del Mar has created a committee to prepare a plan to address coastal flooding predicted to accompany the coming climate change. Although one would think the citizenry would enthusiastically participate in the plan, instead, the plan that emerged has become the problem.

Currently, seawalls and other manmade fortifications offer a partial and short-term remedy. However, such fortifications also exacerbate beach erosion. As was noted in the comments to the draft plan, "seawalls protect property, not beaches or ecosystems." Of the five possible remedial courses of action, the one receiving the most protests involves the moving of structures, sometimes called a "managed retreat approach," in which the relocation of structures is a continuing process. This is not surprising when it is recognized that the implications of this approach are the diminishment or loss of land values. These property values, which were previously very high, would simply disappear. In the end, this option was removed from the plan.[23] Subsequently, the completion of a sea-level rise plan failed to include the state-mandated managed retreat option and means Del Mar may have to forfeit millions of dollars in federal grants. Nevertheless, the senior city planner stated, "managed retreat is not part of this plan and not something we are going to agree on." This is the type of dilemma likely to face property owners up and down the California Coast.

Newport Beach: About seventy miles up the California coast is Newport Beach, a section of which is on Balboa Island, which sits below high tide. Although sea level has risen only nine inches in the past one hundred years, projections of sea-level rise have it rising by six inches by 2035, and nearly two feet by 2050. To protect itself, the city closes floodgates on the island before high tides to keep water from backing up and flooding the island. When tides hit in the future, Balboa Island is expected to be one of the most impacted areas on the West Coast. The city is spending $2 million to add a nine-inch top to its seawall, although this is only a short-term solution. Longer-term residences will be required to put the first floor of new construction some nine feet above mean sea level (the sea level halfway between the mean levels of high and low water). This is about three feet above the current street level. Finally, the whole island is to be jacked up by about two feet by 2050 and another two feet by 2100. It is estimated that seawalls could be elevated enough to protect against up to a five-foot rise in sea level, but the cost would be about $70 million. Part of the solution is to raise many of the homes on the island, although this could cost over $1 million per house.

San Francisco: Still farther up the coast, San Francisco is also recognizing its vulnerability to rising sea levels. Recently, it has proposed a $450 million bond issue to repair and renovate aged sea barriers. Additionally, the California legislature is asking questions about the development of policies that emphasize adaptation to a changing climate. A common concern is maintaining current scenic views, which is difficult if seawalls and houses are being elevated. Also of concern is the loss of property values as the configuration of houses and seawalls change. Given the costs, a common approach is to opt for the short-term solution and defer longer-term responses.

Gulf Coast: The use of dikes, barriers, storage catchments, and pumps can vary greatly when addressing increased sea levels. The dikes in the Netherlands, a country with a long history of the use of water barriers, have both protected lands above the sea and have also been used to protect and reclaim land below sea level. In the United States, Texas is requesting financial assistance from the federal government for a seawall that could stop about 80 percent of the $31.8 billion in damages that Texas A&M University estimates will accrue over the next sixty years as a result of sea levels. The estimated costs are about $12 billion and would involve the building of a seventeen-foot barrier extending sixty miles along the coast. Congress, however, is reluctant to provide such funding since it would create what some believe to be a bad precedent for future federal funding. In other words, if the federal government provided such funding for Texas, it would be obliged to do so for many other states with similar infrastructure concerns.

To meet future storms, New Orleans has built a system of dikes, barriers, catchments, water storage facilities, and pumps, all of which are designed to protect the city, which has large sections below sea level, from both sea-level rise and storm-driven sea surges. It also provides storage for surplus water that passes through the barriers. There is a drainage system consisting of interconnected pipes, pumps, and canals designed to expel the water that passes through the storage system.

Although there are numerous ways to try of adapt to rising sea levels, some have argued that in a situation like New Orleans, the costs of trying to maintain large portions of the city below sea level are simply not justified by the benefits.[24] The optimal policy, it is argued, may simply be to relocate people from low-lying areas of major flood risk to new higher locations with fewer risks. Of course, this is neither easy or cheap. However, in many cases the initial sites may simply disappear as a result of sea-level forces and so would cause considerable loss of life and property.

Houston, another Gulf city, is also regularly hit by hurricanes. Much of the land area is low-lying and subject to flooding. Hurricane Harvey (2017) has been designated, with Katrina, as the worst in history. It damaged somewhere between one hundred thousand and three hundred thousand homes. All of Harris County has required new homes in the floodplain to have a finished floor eighteen inches above the theoretical flood line, while new directives for Houston now require homes to be built two feet above the five-hundred-year floodplain level. Harris County is also developing an additional multiyear flood plan.[25] The voters recently marked the anniversary of Hurricane Harvey by resoundingly approving a $2.5 billion bond proposal aimed at flood mitigation.[26] The regional bond vote, the biggest in the county's history, passed with 85.6 percent of the vote, according to the Harris County Clerk's Office.

A major component of the flood mitigation plan would be a new overflow reservoir estimated to cost up to $500 million. Although Houston already has two flood-control reservoirs, these were overwhelmed by the flood waters associated with Harvey. Thus a third reservoir is under serious consideration. In fact, the local county commissioner has argued that the risks are so great that the plans to build the reservoir ought to move forward immediately, bypassing a time-consuming feasibility study.

The Southeast: According to recent research, coastal real estate and home values in the Southeast region of the United States have experienced substantial losses in value, estimated at around $7 billion, due to concerns over the future impact of sea-level rise.[27] Florida has experienced the greatest losses, but North and South Carolina, Virginia, and Georgia have also been affected.

Despite periodic hurricanes and flooding, Miami continues to experience a construction boom. However, recent studies suggest that real estate market prices are beginning to provide discount properties located in areas of lower elevation and therefore apparent higher flooding risk. Fortunately, major strides have been made in new technologies to make America's critical infrastructure more resistant to natural disasters. New types of construction technologies have been developed and are being applied in Miami. The collective effect of those leaps is to make infrastructure more resilient to both natural and manmade disasters.

Other forms of construction materials and techniques that anticipate future climate change include the renovation and replacement of existing infrastructure with improved models. An example is the replacement bridge from Palm Beach to Mar-a-Largo Island. In anticipation of possible higher waters in the future, the Florida Department of Transportation announced that the new bridge will be four feet higher than the bridge it replaces.

East Coast Resorts: Many parts of the United States' eastern seacoast experience continuing loss of coast line, in part because the continental tectonics are moving inland and to the west. It is common practice in many coastal locations to have annual dredging to replace seashore beaches, which have been eroded over the year and thus contribute to subsidence. Bethany Beach, Delaware, for example, has experienced significant beach loss every winter for decades. Its solution has been a spring dredging and filling effort that renewed the beach for the coming summer tourist season. Within the past few years, it has undertaken a much more comprehensive approach that essentially pushed the beach a hundred yards out into the ocean by creating a huge artificial sand mound. Thus far the approach has been successful. Although this was the city's response to a recurring event, not a major sea-level rise, the future pressures under a rising sea level will undoubtedly be greater.

New England: Like the other coasts, New England has been hit with a series of massive, once-in-a-hundred-years rainstorms. Massachusetts has responded by undertaking a $1.4 billion set of projects for climate adaptation alongside environmental protection programs. The climate projects are focused on seventy-four coastal cities and towns, as well as many inland communities. Much of the funding will be directed at seawalls, culverts, and watersheds, including the repair of seawalls built in the 1930s and 1940s.

COASTAL PROTECTION

The above discussion addresses numerous location-specific cases of threats to coastal land from sea-level rise and surges, as well as the issue of extreme weather events. An aspect of dealing with heavy precipitation and the inundation of heavy rain that can accompany a storm like Harvey or Florence involves the maintenance of large coastal wetlands that have been developed and paved over, resulting in a substantial loss of a natural buffer to soak up water.

A question arises as to how we might anticipate future sea-level rise and invest in adaptive protective measures in advance of that rise. We see an answer in the behavior of towns along the East Coast. The East Coast of the United States is regularly visited by hurricanes. Natural areas along the shore typically provide the ecological service of absorbing the impact of the waves and surplus water volumes, thus protecting properties behind them, including roads and structures. These areas, including sand dunes and wetlands, often experience substantial damage during such events. Other locales have attempted to mimic this response, as in the case of Bethany Beach, Delaware, discussed above.

Sea-level rise associated with climate change will certainly exacerbate damage to such natural infrastructure. Studies show that one-quarter of protected lands in shoreline counties of the East Coast would be affected by a three-foot rise in sea level, endangering lands behind them.[28] Areas especially affected are often those protecting developed towns and cities. The effects would be greatest in Southeastern states. A study of the preparedness of the states, as estimated by their expenditure on investing in coastal protected lands and land conservation, shows variation.[29] Many states invest billions in preserving coastal lands, including parks, wildlife refuges, and other preserves. These state purchases of strategically positioned land parcels enhance their ecosystem protection services, including sea level changes.

FLOODING

Flooding is a risk, not only along sea shores, but along water courses in general. Most climate models predict increased global precipitation. Exactly where this increased precipitation will occur, however, remains to be seen. A recent study[30] used flood physics to produce estimates of flood hazards in the United States. It found that current flood maps underestimated the number of people and asset values located in high-risk flood areas. Over forty million people live in areas where those risks will increase. The study examined the number of people and the asset values at risk based on one-in-fifty-years, one-in-one-hundred-years, and one-in-five-hundred-years peak floods. These numbers did not include the effects of climate change and hence would be substantially greater if climate change increased significantly, as is commonly projected. The findings suggest that future flood problems may be even greater than commonly anticipated if owners ignore the growing evidence of sea-level rise.

WATER DAMAGE BEYOND THE UNITED STATES

Rising sea levels and its accompanying water damage are not limited to the United States. Low-lying regions globally are aware of the likely forthcoming sea-level rises and need to begin planning their long-term adaptation strategies. A study by Newcastle University in England[31] estimated that many European cities would experience the flooding of their rivers. Using climate models, the study ran a number of scenarios analyzing the projected impact for the years 2050 through 2100. Even the most positive scenarios found that 85 percent of cities in the UK and Ireland located next to or near a river would face increased flooding, including London and Dublin. The study also found major flooding likely in northern cities on the Continent, including Paris, Helsinki, and Vilnius. They also analyzed changes in droughts and heat waves for every European city. Southern European cities saw the largest number of heat wave days, while Lisbon, Madrid, and Athens experienced the worst droughts.

The information is now available that new climate issues, including inland flooding, are likely. Countries know, or should know, that current GHG mitigation efforts are unlikely to moderate warming significantly. Thus, they now need to anticipate a warming and resultant flooding damage as possible threats.

Some observers believe that Asian nations have found ways to adapt.[32] In Taipei, Taiwan, for example, hotels are built a few feet from the street, on street level, with ground floors that are mainly concrete.[33] Flood damage is thereby limited. In Hanoi, the Vietnamese built homes with masonry and concrete that don't absorb water like drywall and siding, and therefore don't often require flooded residences to be rebuilt.

OTHER SEA-LEVEL CONSIDERATIONS

A defect in American policy is the subsidization of insurance in the flood-prone coastal areas. These include the National Flood Insurance Program, which aims to reduce the financial impact of flooding on private and public structures.[34] Structures in highly susceptible areas are typically insured, especially those with mortgages. In the US, over five million residences have private flood insurance.[35] However, in less risky locations, many avoid paying for the insurance, thinking they will escape the flood.[36] Thus they are not protected by insurance when unusually widespread flooding occurs.

Although designed to encourage floodplain management, subsidized insurance provides incentives for locating structures in risky, flood-prone locations, since the subsidized insurance shifts a significant portion of the cost and risk from the developer to the tax payer through the federal plan.[37] If a more rapid sea-level rise is anticipated, the insurance premium rates on sea-level properties in a market system should also rise, thereby discouraging new developments in low-lying coastal areas. A subsidy like that of the current federal flood system should be avoided, since it encourages development in high-risk flood areas.[38] Although it has been shown that it is possible to "push back the sea" with the use of dikes, etc., the costs become prohibitive, especially for larger coastal areas within the United States. In the long run, population migration, together with the relocation of infrastructure such as roads, bridges, would need to be undertaken.

The discussion above indicates that there are a host of ways society can adapt to sea-level rise, perhaps the most threatening of the anticipated climate warming dangers. These range from modest dredging and filling operations to multibillion dollar multicontrol systems, such as that employed by New Orleans. Hence, should traditional GHG mitigation fail to stabilize global climate, adaptation provides a viable Plan B. Both mitigation and adaptation assume society will respond to perceived threats. We see evidence of anticipated actions every day in both personal behavior, such a health and person safety, and collective activities, such as national defense and crime control (see the discussion of public goods in Chapter 6).

Nevertheless, issues along the coastal zone raise some difficult legal questions. Several California towns are suing oil giant Exxon for damages associated with present and future damages associated with sea-level rise and climate change, claiming that Exxon knew of the risks and systematically ignored them.[39] Exxon argues that the extent of the possible damages was impossible to know, and they alone were certainly not in error by withholding attempts to provide definitive estimates of future damages. Given the litigious nature of American society, however, such suits will, without doubt, continue to be contested into the foreseeable future.

AGRICULTURAL PRODUCTION

A second area of major concern in terms of substantial future greenhouse damage relates to agricultural production. History is replete with examples of the effect of climate change on food production. In human history, these go back to the biblical tale of Joseph and his family's trek into Egypt, probably about 1700 BCE, in search of food during a great drought throughout the Middle East. More recent experience is found in the United States during the great "dust bowl" of the 1930s in the prairie Midwest states. This event is the basis of John Steinbeck's award-winning book *The Grapes of Wrath* that followed the devastation of an Oklahoma family driven from its farm by the dust bowl and forced to search for a new life in California. Clearly a warmer but drier climate is generally not conducive to higher productivity for most agricultural crops in their current locations. However, agricultural crops can be relocated.

In an RFF study,[40] we used agricultural production and forecasting models to compare the impact of a reoccurrence of a dust bowl-like climate change on current agricultural productivity of four states of the Midwest: Missouri, Iowa, Nebraska, and Kansas. The findings indicated that production success was highly dependent upon the farmers' behavior. If farmers' practices remained stagnant in the face of a changing climate (called the "dumb farmer" scenario), the study suggested there would be large declines in productivity, crop output, and incomes, declines similar to those of the 1930s. However, if farmers adapted their behavior and agricultural practices to the new climate (if they were smart farmers), much of the damage could be avoided or circumvented. The type of changes required involved, most importantly, changing to more suitable crops conducive to the existing climate. For example, a grain crop for the next season could be changed from soybean to wheat. In many cases, such a radical change to an entirely different grain might not be necessary. The farmer might simply change from one type of wheat (red winter wheat) to another type (white wheat) with a somewhat different genetic composition more suitable to the new climate. For most crops, there are a wide array of provenances, or lineages, and a range of genetic variation, whereby seed of the appropriate crop lineage can thrive under various conditions. Careful selection of crop and seed type can mitigate much of the loss of productivity that would likely occur if the crop and seed type were left unchanged.

Modern genetics has provided certain customized seed that has been genetically modified either by traditional breading technics or by modern asexual genetic modification. These modified seeds (germ plasma) offer characteristics that are designed to flourish under alternative, often harsh conditions. For example, drought resistant wheat might be substituted for high-water-demanding corn. Thus, genetically modified seed more suitable to the new climate environment could be substituted for traditional seed. The similarities, which often stay within a certain species type, are likely to make the effective biological substitute an acceptable change in the market, like oats replacing corn as feed grain or rice substituting for wheat for pasta. Such alternatives were not readily available in the dust bowl period but are commonly available today.

Although selection of crops with varied genetic bases can address some of the changes likely to be associated with climate change, this response will not always be adequate. Indeed, more extreme measures may be needed. For example, a farmer might convert a grain field to an entirely different use—from wheat to pasture or orchard[41]—with an eye to market acceptability. Finally, there are other naturally occurring potentially favorable factors. Changing temperatures need not always be adverse. In some locations, a modest temperature increase will increase the productivity of many agricultural crops, including some that may not have been feasible under earlier cooler conditions. This surely was true for the Vikings in Greenland and would likely be true in much of Nordic Europe today, as well as parts of the United States and Canada. Warmer weather, if accompanied by adequate precipitation, benefit many types of agriculture in many locations. Countries in cooler northerly regions have sometimes responded to the prospect of global warming with glee, noting the new growing opportunities that the warming weather could provide.[42]

FORESTS AND ECOLOGICAL DAMAGE

Forests play an important role in human life. In addition to providing a host of goods and services, global forests are an integral part of both the global ecological and global carbon systems. We think of forests as providing wildlife habitat, clean water, air filtering, and recreation, as well as timber production for construction and paper. But forest growth involves carbon dioxide being captured from the air, with the carbon molecule being used to form wood cells of the plant while the oxygen molecule is released back into the atmosphere. Global forests thus hold captive large volumes of carbon, as well as having the potential to capture additional volumes of carbon from the atmosphere. Humans can use the carbon-capturing feature to enlist forests in both the mitigation and the adaptation process.

However, forest destruction can release carbon, and deforestation is a well-recognized cause of the buildup of carbon dioxide in the atmosphere and the resulting global warming. Concerns over the state of global forests and their contribution to atmospheric carbon have been expressed over the past several decades. A recent study estimated that 10 percent of the world's carbon dioxide emissions are from tropical deforestation.[43] Although evidence has developed that temperate forest deforestation has been stabilized, concerns about tropical forest depletion remain high.[44]

Concern is often expressed about the effects of warming on forests. Early assessments sometimes portrayed forest dieback anticipated from climate change to be massive and pervasive, leaving behind a desolate moonscape. Such a moonscape outcome is now viewed as unlikely, since a dieback of current forest is likely to be replaced rather quickly by more suitable species that are likely to move into vacated spaces. The more extreme dieback might only be expected if warmer temperatures were very rapidly imposed on the system, especially if the warming were accompanied by a substantial decline in precipitation. Under these unique conditions, the rate of dieback would exceed the rate of forest renewal. More generally, decreased moisture would normally change the composition of a forest, perhaps even more dramatically than the effect of a warming. Precipitation decreases, if large enough, would typically create changes where, over time, grasslands would replace a forest.

Damage associated with climate impacts on forests can be of two types: commodity and environmental. Commodity losses in terms of the loss of industrial wood for construction and paper are likely to be small, in that a high percentage of industrial wood comes from planted forests. These are typically harvested before they become so mature that they are likely to be vulnerable to wildfire. For planted forests, adjustments can be made as to the species of tree and other characteristics to render their genetics consistent with the changed climate.

A type of response to the damage caused to mangrove forests in the country of Myanmar has been suggested after being observed by drones. Using Lidar technology, which can provide images of the topography of the land even if it is covered by vegetation and trees, the drones can provide a 3D map of the land surface and measure soil type, quality, and moisture. Using this information, a second drone can shoot biodegradable seedpods while flying just above the ground. Such an approach could be applied broadly at very low cost, adjusting to local conditions, to repair the damage to different types of forest.[45]

Most studies now suggest that in the long term, climate change will have an overall neutral or slightly positive effect on forest extent and growth globally. In some areas, forests will be advantaged, while in others they will be disadvantaged, with the total global area suitable to forest only modestly changed.[46] This result is expected partly due to the inherent adaptability of forests, and also due in part to the fertilizing effect of CO_2. The increase will also be due to the general overall rise in global precipitation. A question that remains is the mobility of forests in response to climate change. Trees have mobility over the long term. They are capable of relocating to take advantage of a more suitable climate. Examples of this are found on highlands and mountains where species can, and have, moved up and down the mountain as the climatic conditions change. There is also the experience of wild forests gradually relocating in response to changing conditions. However, not all trees have the same mobility and a moving forest may change its composition of species as it responds to climatic change.[47] A major adaptive action that humans could take would be that of facilitating tree adaptation and mobility. Seeding, planting, and providing for the natural seed mobility of plant species to new, more suitable areas would be a desirable activity both in promoting future industrial wood and also in promoting future forests to provide environmental benefits.

Forests would not all respond the same as they react to global climate change. Decline would be experienced in some places while expansion might occur elsewhere. Overall, studies have concluded that unconstrained forests would experience modest positive expansion. For example, Brent Sohngen of Ohio State University and Robert Mendelsohn of Yale show that expanded forests in some areas would be expected to offset the negative climate effects on forests elsewhere and project that the total forest area would expand globally.[48]

Although the overall effect of climate warming on forests is likely to be small, damage that occurs in some regions could be significant, as could the damage caused during transition periods. Humans can adapt forests to these damages in a number of ways. First, forests can be harvested before the dieback is complete, thereby removing valuable industrial wood. Additionally, salvage logging is common, and much of the value of the trees can be captured even after they are dead. Finally, humans can intervene to promote rapid regeneration and reforestation. Useful activities would be to introduce seeds and/ or seedlings appropriate to the new climate into the forest in a timely fashion. Broad widespread reforestation might even involve cheap aerial seeding.

Another element of the land-use question is that of the protection of wildlife and other noneconomic assets of the forest that might require some type of human intervention. Animals are very mobile, and some species of trees also exhibit a significant amount of mobility, especially when their environment changes substantially. Nevertheless, human management would need to be aware of these issues.

VEGETATION CHANGES AND WILDFIRES

Another important issue is forest wildfire brought on by natural causes like lightning strikes. Forests are most susceptible to fire when they are old and under stress. These situations might involve disease and infestation. A changing climate undoubtedly will stress many forests as the change puts trees out of their optimal biological range for growth. This would threaten the health of the trees and perhaps the entire forest. Thus, a major concern with climate change is that of managing an increasing number of forest wildfires. The optimal strategy is probably not to attempt to eliminate the fire, but rather to control the burning to minimize ancillary damage and to facilitate the forest's transition to a forest type more suitable to the emerging climate.

The process of warming can be disruptive to forests and any adaptation strategy must anticipate this issue. As the climate changes, some trees will become located in areas where they do not flourish. These portions of the forests are likely to experience dieback, thereby releasing some of their carbon from decomposition and/or fire. As with other natural systems, the life cycle of the forest includes birth, life, and death. The cycles of individual trees are often different from that of the forest system. Individual trees come and go while the forest continues. But, as a whole, forests may change the composition of their trees: for example, they may shift over time from pine to fir. These overarching compositional changes are more likely with major climate change.

Wildfire is likely to occur particularly in temperate and northerly forests, where it is common today. Global warming will undoubtedly change the mix of temperature and precipitation in such a manner as to encourage changes in forests as they adapt to the climate change. Some of these changes are captured in regional General Circulation Models (GCMs) that project the expected change in natural vegetation likely to be associated with the changing climate of a region.[49] GCMs are numerical models that represent physical weather processes and projected changes in the global system by region, including the atmosphere and oceans, for simulating the response of the global climate system to increasing GHG concentrations. Although not all GCMs agree on the precise climate and ecological changes for each individual region, there is agreement that regional conditions will change and with them the nature and type of vegetation and forests. Some species of trees, like pine, and other vegetation will be ecologically advantaged, while others, such as deciduous trees, may be disadvantaged. This will necessitate a transition in the mix of tree and plant species over time. Dieback will occur for disadvantaged species while others are advantaged and will expand. Applications of GCM will be discussed later in this chapter.

Species experiencing dieback will create conditions conducive to wildfire. These include fuel buildup in the form of dead trees and vegetative materials due to disease and infestation. The wildfires will follow and clear the forest area, opening it to invasion by different species, some of which will be more suited to the changing conditions that prevail. Widespread global climate change will promote the concurrent transition of other forests and their ecosystems. New ecosystem types will emerge as species previously not commonly found together may find new conditions suitable to both.

For many types of tree species, like pine, fire plays a major role in their ecology. Pine are shade intolerant and will flourish in newly cleared spaces that experience intensive sunlight, such as those spaces created by widespread fires. In general, pine will grow en masse as an even-aged forest,[50] because late arrivals to a pine forest will not flourish after earlier arrivals generate shade that is not conducive to the growth of later pine seed and seedlings.[51] In the longer term, the forest composition will change depending upon the mix of climate conditions, seed sources, and other disturbances, including fire.

In the adaptation process, humans can anticipate forest changes driven by climate change and may undertake programs designed to facilitate the transition process. The process of wildfires contributes to the release of carbon dioxide into the atmosphere.[52] A balanced management of these processes is called for. In planted forests, the species mix can be adjusted by providing seedlings chosen because they are likely to be successful in the changing climate conditions. In natural forests, plant life managers can make sure that adequate seeds are provided on an appropriate time scale, even by aerial seeding.

Biodiversity loss has been a concern voiced well before the concern arose about climate change. Early issues centered around the extinction of specific plant or animal species. Losses in the United States, for example, have included the Passenger Pigeon and the American Chestnut tree. Near losses have occurred with the American bison, various birds, such as the western sage hen, as well as some species of whales. Many of these losses are attributable to the destruction of habitat due to land-use change.

Various approaches have been undertaken to assist in the preservation of threatened species and biodiversity generally. One approach has been to attempt to maintain a dynamic forest when disturbances are followed by tree and species replacement. The establishment of seed banks, which are often geared primarily to food crops but may also include other plants, especially some that might be threatened, allow for disturbed areas to be assisted by the introduction of appropriate seeds. A common approach to biodiversity preservation has been to establish protected areas that often have multiple uses but also serve as biodiversity reserves, preserving plants, arthropods, and animals. One function of the United States National Park System, and that of many wilderness park systems worldwide, is that of a biodiversity reserve.

A change in the nature and distribution of the earth's climate would no doubt put additional pressure on biodiverse areas around the globe and thereby complicate remedial actions. Climate change would precipitate land-use changes and an associated reallocation of plant and animal species. Some have claimed that climate change is expected to threaten the extinction of up to one-quarter of all species on land by the year 2050. A study by the World Wildlife Fund and others estimated that the UN forecasted temperature rise of 3.2 degrees Celsius by 2100 will render many habitats unsuitable for indigenous plants and animals, resulting in widespread extinction.[53]

An important feature in preservation would be the development of a flexible system that would facilitate the mobility of species. Note that land-use changes driven by climate change are likely to be relatively slow and therefore would allow time for many of the species to relocate. An important feature of any flexible system is the connection of habitats. In this case, forests and habitats should be connected in such a way as to minimize isolated pockets and promote connecting corridors to facilitate mobility among isolated populations. Efforts to identify promising areas are already being undertaken for current conditions, including Torres del Paine in Chile and Banff Wildlife Bridges in Canada.[54] Additional work anticipating future conditions under major climate change could be instigated. The use of General Circulation Models for the geographic areas under consideration would be useful.

GENERAL CIRCULATION MODELS

GCMs are likely to find increasing uses as climate change persists and management and adaptation to climate-related damage becomes increasingly important. Such numerical models of a region's physical processes are usually scaled down to simulate and project the possible impacts of future climate changes on the region's climate.

GCMs offer the potential of being very useful in assisting climate managers in anticipating future damages and planning for current and future remedial actions. They can be particularly useful in making decisions in agricultural, forestry, and other land uses by providing simulations of temperature and precipitation in specific locations within and across a region.[55]

A current problem with GCMs is that several alternative models are available for most of the regions. This reflects different assumptions as to the nature and timing of the physical processes. Not all of these generate consistent simulations of future conditions. For example, many regional models of the US South have anticipated increasing drought as climate warming progresses, gradually eliminating the southern pine forests. However, others do not project that result. This raises the question of which model to use, especially when the projections are quite different. Most approaches have chosen a hybrid that represents an average of the models with the most common outcome, even if simply choosing the model with the most common average is no guarantee of correctness. These models are likely to be useful at minimum in identifying the components of the regional system that require special attention.

EXTREME WEATHER EVENTS

The relationship between climate change and extreme weather events remains an open question. Does climate change promote additional hurricanes, change the magnitude of the hurricane, both, or neither? A case in point is the incidence of large tropical hurricanes such as Florence. It is generally agreed that in a warmer world, precipitation will increase. Warmer waters can result in an increased amount of moisture in the atmosphere. Studies have estimated the increased precipitation associated with warming to be 1.5 to 3.0 percent for each degree Celsius increase in global temperature.[56] Some believe that climate change may increase the intensity of hurricanes, while not necessarily increasing their frequency. But the link between climate change and hurricanes remains unsettled.[57]

Adaptation to climate change, however, necessarily involves a readiness to deal with hurricanes and their surges, particularly if they became larger and/or more frequent. Roger Pielke of the University of Colorado suggests three approaches to anticipate and address large climate disasters. First, the establishment of disaster review boards to examine the responses and allow for learning to improve future responses. Second, the encouragement, through regulations and incentives, of resilient growth, in which flexibility in growth and response is critical. Third, for the United States, an enhanced federal response capacity is needed.

A series of recent studies by the National Center for Atmospheric Research (NCAR)[58] predict that there will be mostly financial benefits from mitigating climate change—that is, damages will be reduced. But some risks are unlikely to be reduced, including the financial exposure from future very large storms resembling Hurricanes Harvey, Irma, and Florence. Better insurance coverage, better building codes, and encouraging relocation to more secure geographic regions that are less susceptible would be a way of making reductions in financial exposure.

A summary of these studies warns of a nonlinear relationship between the strength of future storms and potential financial losses, which stand to increase even if the worst predictions about hurricanes aren't realized. Financial exposure could increase as much as five times faster, because the number of buildings and the population in coastal areas continue to proliferate. So, as will be discussed, changes such as charging more for insurance (by eliminating subsidies), requiring the building of structures to be much more secure, and offering incentives to those who choose to live in more secure areas while charging a risk premium to those who live in more danger-prone areas would address some of these issues. Not all can benefit directly, but the general trend will be to avoid high risk areas where possible. State and local governments can reduce their risk by adopting stricter zoning laws and building codes.

NCAR studies, collectively titled *Benefits of Reduced Anthropogenic Climate Change*, are an effort to show the benefits and the costs of global efforts to reduce greenhouse gases. The studies are intended to mesh together science and other disciplines, such as insurance, that measure some of the political and economic impacts of combating climate change. They show that when it comes to reducing extreme heat, for example, the benefits of lowering greenhouse gas emissions are clear and substantial. In other areas, such as in agriculture, where more CO_2 may help fertilize crops while also causing warming-related crop damage, there are "stubborn uncertainties" that make benefits harder to predict.

In the future, more accurate risk models will be needed to back up climate models, ones that ultimately show the accumulating dangers of consolidating populations on coastal shores. The potential of stronger storms is only one element. Sea-level rise by itself makes flooding worse, and the use of ground water, which promotes sinking land, exacerbates this problem.

RECENT EXPERIENCES AND SOME LESSONS FOR FUTURE PREPAREDNESS

Preparing for the Future: Puerto Rico

A well-known politician famously stated that huge crises also create huge opportunities.[59] In the case of Puerto Rico, a huge opportunity has been created from Hurricane Maria. In September 2017, Hurricane Maria tore through Puerto Rico, only two days after Hurricane Irma, bisecting the entire island, bringing 150 mph winds and torrential rains to some of its most populated areas. It had been estimated by the Harvard School of Public Health that over 4,000 people died as a result of the hurricane and associated damages,[60] although the final count was still a massive 2,975.[61] Puerto Rico's entire power grid was knocked offline during the storm. At the time, reports suggested that it could be four to six months before power was restored on the island, while the Corps of Engineers estimated it would take over eight months. Both of these estimates turned out to be optimistic. It took even longer. It was well over half a year before most of the island's 3.4 million people received grid power, and many had not received power until a full year had passed. By that time, many hundreds of thousands of people had left Puerto Rico, some perhaps forever.

Top lawmakers have said that the unprecedented destruction provides a rare opportunity to completely rebuild a modern electric grid from the ground up. In the case of Puerto Rico, the electrical grid was very old, in poor condition, and in need of major repair or replacement.[62] One reason for the poor system was poor governance. For decades, many of Puerto Rico's municipalities received free power, and many others were billed but didn't pay. There were few funds being collected and fewer being spent to maintain the system.

In short, Puerto Rico was a social and economic "basket case" before Maria arrived. It relied extensively on debt financing. Public systems were dilapidated and not well maintained. Government was excessively bloated and bureaucratic. Unemployment was very high, with one in five workers unemployed. Puerto Rico was also in financial default, having stopped paying bondholders in 2016.[63]

At a congressional hearing, concerns quickly focused on the power facilities. One congressional staffer said, "At the very minimum, (power) generation needs to be privatized, and we need a very strong, independent regulator that ensures fees are being charged fairly and everyone is paying."[64] However, given US regulations, the issue of reconstruction came with some legal problems. FEMA showed that it was willing to provide funding for rebuilding, but it demonstrated the limitations on the improvements it was to finance as instructed in existing legislation. Congress has shown that it will support legislation to provide a waiver from current funding limitation, and the president authorized additional funding support.[65]

To meet future storm challenges that will surely come with or without the impacts of climate change, a power system is needed to replace the destroyed electrical system, one that provides flexibility and resiliency. Decentralization (several electric power generators distributed throughout the system) is probably a critical component of any such system. An example would be one like the system provided in Alaska, with a decentralized approach that combines renewables, fossil fuels, and storage. Such an approach may be useful for similar "islanded" grids in the territories.

One innovation being introduced by a private company, Sistine Solar, is that of rooftop solar energy.[66] Earlier installations were hooked into the grid and these customers lost power like everybody else with systems tied to that same grid. The new installations will come with battery storage to keep the systems running even during hurricanes and provide for solar recharge.

Another similar proposal is that of a system consisting largely of micro-grids. A micro-grid is a segment of a larger electrical grid that operates to some extent on its own power sources, which are chiefly fed by renewable energy, and can detach from the larger grid in cases of emergencies. Such a system, while technically attractive, would need to successfully navigate Puerto Rico's intractable political landscape.

Considerable thought is now being given to the possibility of installing renewable energy capacity,[67] particularly solar power and storage systems, to the island. Several companies have provided renewable solutions. Solar power offers a decentralized solution so that future storms cannot destroy the entire system by simply disrupting a central power plant and its power lines.

Some in Congress favor a build-it-better approach. In a floor speech two months after Hurricane Maria, Senator Lisa Murkowski of Alaska talked about working with the Army Corps and the Department of Energy (DOE) to reshape Puerto Rico's grid for greater resilience. "We want to look at not only the damage caused, where recovery efforts stand, but also lessons learned," she said. The senator talked about building a more resilient and sustainable grid.[68]

At the DOE, some officials have expressed interest in tackling elements of the rebuilding task as "demonstration projects" to bring in newer technologies. Outside of the confines of Washington, Silicon Valley's big clean energy technology names, such as Tesla Inc. and Sunrun Inc., are joining makers of large-scale batteries such as AES Corp. to pitch their technology to Puerto Rico.[69]

Some members of Congress have focused on reshaping the legislation for the Puerto Rico Electrical Power Authority (PREPA). The director of PREPA was forced out as a result of his inadequate responses to addressing the problems due to the storm, and attempts have been made to shape legislation to guide the debt-ridden commonwealth back to solvency.[70]

Others in Congress have looked for ways to make the oversight board stronger and more independent of political forces. Governor Ricardo Rosselló moved to break up and privatize the power authority. This change required reopening the legislation to fine-tune its oversight provisions, according to committee staff.[71]

Similar broad changes to promote flexibility and resiliency that could be useful for anticipating and addressing future climate change are under consideration. Legislative options for Puerto Rico might be to permanently exempt the island from the shipping restrictions imposed by the 1920s-era Jones Act.[72] This restriction, which requires all shipping from the United States to be on American produced ships,[73] inhibited the delivery of critical materials and supplies by non-American carriers in the wake of the hurricane.

A related hurdle is Puerto's Rico's ongoing fiscal crisis, much of it driven by the Puerto Rico Electric Power Authority's nine-billion-dollar debt. Other issues associated with the hurricane include cell service, which was out on almost the entire island, in part due to the absence of fuel for backup generators to provide battery power. Other problems persist, including an ominous warning about the Guajataca Dam, which is near the point of breaking, threatening downstream areas with deadly floods as well as additional power losses. Overall, getting the power back on in Puerto Rico has been a daunting and expensive task. Transformers, poles, and power lines twist from coastal areas across hard-to-access mountains. In some remote areas, the poles have to be maneuvered into place with the help of helicopters.

Puerto Rico is said to be located in hurricane alley, so it is a bull's-eye for storms. Given this characterization, it is remarkable that it seems to have been so totally unprepared for hurricanes like Maria. It is hard to imagine how heavily Puerto Rico would be damaged if a huge hurricane were to strike after climate change had raised the sea level only another foot. Given the rising GHGs and increasing temperatures, it appears likely that Puerto Rico will be revisited by massive storms in the not-too-distant future.

Although FEMA believed it was ready in advance to address the damages associated with Hurricane Maria, this optimism was obviously misplaced. Even though supplies of diesel fuel were gathered in anticipation of powering the generators of hospitals, refrigerated facilities, and other public works, the damage to the transport infrastructure and the absence of gasoline precluded timely distribution of transport fuels. Although there were stockpiles of critical materials in storage at the ports, the same limitation in transport and communications severely curbed their timely distribution to areas of need. Archaic maritime laws, such as the Jones Act, limited the opportune delivery of critical material to Puerto Rico from the outside.

Out of all this, there is good news and bad. The good news is that we have learned a great deal about the nature of damage associated with a massive storm, and this should help us prepare for future disasters. The bad news is that although it was believed that the preparation was adequate, it clearly was not. Backup systems are not useful if distribution and communication systems are knocked out. Fuel is not useful if it cannot be delivered to critical locations. As a result of the storm, we had a region with an old, poorly maintained infrastructure that was heavily damaged, and a power system that was almost destroyed. This situation forces replacements, presumably with more modern and resilient systems. The decentralization of power needs, for example, suggests a more diversified system that contains localized, independent units. One can only hope that the lessons learned are being implemented.

But rebuilding the system requires financing. Since Puerto Rico had declared bankruptcy shortly before Maria hit, the adequacy of its financial situation has seemed bleak. Even if it were solvent, much of the recovery funding would need to come from external, mostly federal, sources. Fortunately, in the wake of the three hurricane disasters, the federal government's earlier appropriation of funds through disaster relief programs can be used in the recovery and with flexibility to institute sensible innovations.

Additionally, the situation in private financial markets may not be as dour as it had seemed. The *Wall Street Journal*[74] paints a somewhat more positive future financial picture for Puerto Rico. It notes that the island's bonds have made a strong recovery following their initial precipitous decline both before and following the hurricane. The recovery occurred despite the fact that the preliminary reaction to suggestions that the island's debts be forgiven was quite negative.[75] The journal also noted that the Puerto Rican domestic economy has responded quite favorably to the rebuilding boom following the storm. The rebuilding should involve systems with greater resistance to withstand the more severe storms that will undoubtedly descend.

SOME INFLUENCES OF ATMOSPHERIC GREENHOUSE GAS

Greenhouse Fertilization: Although GHG and especially carbon dioxide are the sources of destructive warming, they can also bring some benefits to humans. Evidence suggests that in many places throughout the world a range of temperature increases of up to 2 degrees Celsius could be positive overall for agriculture.[76] Climate research suggests that global warming will be associated with greater global precipitation. A wetter climate together with a warmer environment will be positive for agricultural in many areas. Finally, carbon dioxide can promote biological growth in many plants enhancing agricultural productivity.[77]

Another concern that was touched on above is that of the effect of increased CO_2 on plant growth. Although this seems to be a controversial issue in some circles, the science is pretty clear. It is well-known that increase CO_2 has a fertilizing effect on many types of plants such as corn and cactus,[78] thereby promoting more rapid biological growth in warmer environments. Enhanced CO_2 is a technique used in many plant nurseries to accelerate the growth of certain plant seedlings. This is sometimes referred to as the CO_2 fertilization effect.

The MINK Climate Study, an RFF study, brings up another consideration.[79] It notes that in increasingly arid conditions it is sensible to introduce irrigation that might not have been justified earlier when natural precipitation was more prevalent, subject to the availability of water. It is the introduction of external water that has been the source of the high productivity enjoyed by many midcontinental regions in the United States.

In a global context, major changes in climate across regions can restructure the geographic distributional pattern of agricultural production. This would necessitate adjustment in subsequent agricultural product trading patterns. These changes, although perhaps initially disruptive, would likely be adjusted by trading markets. There may be a potential problem if food transport hubs, such as Kansas City, Missouri, became choke points. Railroads, aging canal locks, and other infrastructure absence or deterioration could create transport problems. Obviously, successful adaptation in the food distribution area requires an appropriate and adequate system of transport that would adjust to changing production patterns.

Another agricultural production problem that may be associated with climate change is the change in water availability from glaciers and snow melt. California snowpack and melt is critical for both agriculture and human consumption. In much of the rest of the world, runoff from glacial melt is the source of much of the agricultural water. Peru, for example, relies heavily on runoff from melting glaciers, as does Nepal and parts of India. Declining glacier mass and runoff would threaten agriculture production in many places.

Land-Use Adaptation for Capturing CO_2: An additional aspect of adaptation is found in practices not directly related to the management of damage, but rather in attempts to reduce atmospheric levels of CO_2 after it has been released into the atmosphere. The Global Carbon Cycle promotes the ready flow of carbon among the biosphere, including the atmosphere, and the oceans (as well as, with more rigidity, through the lithosphere) (see Chapter 2). Large amounts of carbon are found in the biosphere, particularly in forests and soils, and can readily shift back and forth between the biosphere and the atmosphere. The biosphere can function as a significant sink (absorber of GHGs from the atmosphere) as well as a carbon source. Carbon is also readily captured by the oceans, but in contrast to carbon in the atmosphere, its transfer back to the biosphere from the oceans is less fluid than the transfer from biosphere to atmosphere.

Forests are commonly managed to promote their growth and

expansion for financial purposes, such as the production of industrial wood such as lumber and pulpwood. An additional rationale for forest production is their function in the preservation of the natural environment and the multiple services they provide, such as watershed protection, erosion control, wildlife habitat, and biodiversity protection. In recent years, we've begun to appreciate that forests function as a vehicle for capturing and storing carbon from the atmosphere. The carbon sequestrating function of forests, which absorb CO_2 by way of photosynthesis thereby lowering the level of GHGs in the atmosphere, provides a potential for reducing the pressures of additional climate warming. Thus, the use of forests for carbon capture and subsequent storage provides another tool to promote adaptive management.

BIOENERGY AND BIOMASS

A renewable energy can be viewed as a mitigating agent in that it replaces fossil fuels. In its capacity as a substitute for fossil fuels it can also be viewed as part of the adaptation process. Biogenic energy is of significant importance, but not discussed nearly as often as other renewables such as solar or wind. Biogenic energy involves energy gleaned from using biological materials, predominantly wood and wood waste. The energy would involve the direct burning of biomass, but also includes the use biofuels such as ethanol, commonly used with gasoline, or methanol, produced from biological materials. Ethanol and methanol are liquid fuels commonly made from grains, although they can also be produced from wood, other vegetative matter, and even animal waste. Some of these materials are used as substitutes for fuel oil. About 10 percent of US energy is from renewables, of which the largest portion is bioenergy, at about 5 percent.[80] A large portion of bioenergy comes from the use of biomass waste to power certain industrial applications, as in the wood processing industry that uses waste wood for power. Some studies indicate that biomass constitutes around 10 percent of the world's total energy;[81] much of this is used in the developing world.

Biomass is viewed as a renewable resource since regrowth can renew it, either natural or artificial. Unlike other renewables, burning biomass releases GHGs directly into the atmosphere. If not used for energy, biomass will typically decompose over time, and the decomposition will also eventually release GHGs into the atmosphere. Biomass and biogenic materials can be made renewable if the bio-source from which they are drawn is regenerated as new growth. In this context, no net emissions are created, and the system is a zero-net carbon emitter over the long-term cycle. The process is this: growth, which captures carbon dioxide; burning for energy purposes, which releases carbon; and regrowth, which recaptures the carbon. Thus, the net effect over the cycle is zero emissions.

GEOENGINEERING

The term geoengineering includes a host of technologies, many directed at the atmosphere, designed to make atmospheric warming neutral even in the face of increased GHGs. Other geoengineering may involve changing the reflectivity of Earth's surface and capturing carbon and carbon dioxide in the ocean bottom or in permanent underground storage chambers. These technologies are another aspect of adaptation and discussed in the following chapter.

BIOLOGICAL SEQUESTRATION

Adaptation can also be accomplished by changing the reflectivity of the atmosphere or by changing the reflectivity of the earth's surface through land-use management. Much work has been done looking at the potential of forests to sequester carbon. The basic idea is not to prevent the emission of GHGs, but to provide a vehicle to capture them once they have been emitted. Biological growth, especially of trees, offers the ability to sequester carbon. Trees are long-lived and have the ability not only to capture carbon but to store it in cells for very long periods. Thus trees, and the forest in general, are able to accumulate large volumes of biomass, which consists of large amounts of carbon. As noted earlier, the process of growth requires carbon dioxide, which is drawn from the air. The carbon molecule C, of CO_2, is captured in the tree wood cells. Simultaneously, the oxygen molecule, O_2, is released as oxygen into the atmosphere. This release of oxygen has sometimes been used as the rationale for calling forests the "lungs of the planet." Studies have suggested that an expansion of Earth's forests could capture enough carbon to offset up to an entire year's total release of carbon dioxide emissions[82] It has been suggested that tree planting, although not an antidote to warming, might "buy time" for the institutionalization of more permanent carbon stabilization solutions.

Part of the process of adaptation might be the introduction of systems to promote forests' carbon capture. These include forest carbon credits or carbon offsets, where payments are made to forest owners for increasing their volume of trees and other plants. This would require a broad policy whereby payments (taxes) would be required for carbon emissions that are then offset with carbon credits, obtained by undertaking activities that capture carbon. Carbon credits might be earned by establishing new forests. Such a system has not yet been broadly or successfully established. Attempts have been made to introduce forest carbon credits to market trading commodity exchanges (such as the Chicago Mercantile Exchange), to be traded to polluting entities. However, these attempts have not yet been very successful due in part to forest-monitoring problems on a global scale and the lack of market demand willing to make financial payments.

In addition to forests, soils can capture and retain carbon. Certain types of agricultural practices, like "no-till" cropping certain grain crops, allow the soils to build up their stock of carbon. This provides two services: it improves the soils for future agriculture, and it provides another vehicle for removing carbon dioxide from the atmosphere. Vegetation, and particularly grasses, has the capability to promote the buildup of carbon in soils. By capturing and retaining carbon via the growth of grasses and the decay of plant life, including roots to organic matter, the buildup of carbon can be achieved and stored in the soil.

Thus possibilities exist for adaption activities that offset atmospheric carbon buildup, thereby contributing to the adaptation process and achieved through land management practices.

CAP AND TRADE

A variation of forest carbon credits is gaining headway in the electrical power sector in parts of New England. A program called the Regional Greenhouse Gas Initiative involves several northeastern states. California has its own similar program covering not only power facilities, but also factories and oil refineries. The idea is to have power companies in the region with power plants required to purchase permits from the state for the carbon dioxide they release. This is a way, like a tax, to make the polluters pay for their carbon emissions. The state also earns revenue from the permits that can be directed toward climate issues. The permits constitute credits and can be sold or traded to other companies. However, the number of permits controlled by the state will diminish over time, putting increased pressures on the power companies to lower their emissions. The system is sometimes called "cap and trade."[83]

CLEANER FOSSIL FUELS

A bright spot in the mitigation battle is the decline in the total amount of coal used as a source of energy. Natural gas has increased, taking some of coal's share in energy production. This is desirable from a GHG perspective since natural gas releases only about half of the GHGs per unit of energy as coal does. This shift to natural gas results in part from a desire to reduce GHG emissions but is also due to the development of new technologies, especially hydraulic fracturing (fracking), which has resulted in significantly cheaper natural gas, as well as new sources of petroleum.[84] However, despite cost advantages, fracking is controversial due to its sometimes-negative effect on water quality and other environmental damages. Natural gas has now replaced coal in many applications and locations. From the GHG perspective, lower gas prices are a mixed blessing. On the one hand, lower prices incentivize the substitution of low carbon emissions natural gas for high emissions coal. On the other hand, low natural gas and petroleum prices discourage the substitution of renewables for fossil fuels.

From the financial perspective, the United States has benefited greatly from the development of fracking to obtain natural gas and petroleum at competitive costs. The availability of these new low-cost energy sources has benefits in providing relatively lower energy power costs. The United States has become a significant exporter of natural gas and, at the same time, reduced its requirement for imported petroleum substantially. Although the United States did not sign the Kyoto Protocol nor agree to its predetermined GHG reduction targets, the widespread use of natural gas has, according to then Secretary of State John Kerry, allowed America to meet the emissions reduction target for which it would have been assigned under Kyoto.[85]

CARBON CAPTURE AND STORAGE

Another fundamentally mitigative approach could allow for the continued use of fossil fuels. Clean fossil fuels could continue to be used if the carbon were somehow removed before it was allowed to escape into the atmosphere. Note that the ultimate problem is not fossil fuels, per se, but the climate implications of carbon released as carbon dioxide into the atmosphere upon the burning of fossil fuels. An example of the use of fossil fuels while at the same time separating and capturing the carbon is found in the process called Carbon Capture and Storage (CCS). This is a technology that diverts the waste gases after the energy has been utilized and then captures the carbon from the gases before those gases and the carbon are released into the atmosphere.

The usual case of CCS involves the diversion of the carbon dioxide from the gaseous waste stream, which would otherwise be released into the atmosphere. The carbon dioxide is separated from the other gaseous streams and routed for storage, usually in permanent underground storage facilities, such as geological formations including former oil basins. In this way, fossil fuel can continue to be utilized for energy but without the negative effect of increasing atmospheric GHGs. Carbon injections into geological oil formations have been used for some time to increase the yields of older operating wells.[86] More recently, prototype facilities have used this technique to produce power facilities that are free of carbon emissions. Such facilities have been used operationally in the North Sea and Canada. Typically, these facilities redirect the captured carbon dioxide into subsurface formations. Although the CCS technology is a sound approach for allowing the continued utilization of coal, the economics are questionable compared to the costs of solar, wind, and particularly natural gas, all low carbon-emitting energy substitutes for coal that can be obtained easily. Clearly the problem with CCS is its cost. The process increases costs of coal power by an estimated 50 percent, making the cost of coal even more noncompetitive with natural gas and other energy sources.

One of the questions is: can anything useful be done with the carbon dioxide captured from a process such as CCS? As noted, in many instances the gas is simply to be injected into an appropriate geological formation. In some cases CO_2 is injected into an oil formation to help force out remaining oil and/or gas. In others, it is simply left in the formation forever. However, recent research suggest the possibilities of a lithium/carbon dioxide battery. In this case, the CO_2 could have a useful life after it was removed to form a fossil fuel.[87] Although very attractive, the operationalization of such an innovation appears some distance in the future.

CONCRETE CAPTURE

One of the major sources for GHGs is the production of cement. Cement is a major component of concrete, which is used worldwide in construction. Cement turns limestone into lime by baking at high temperatures and thus produces copious amounts of CO_2.

Although cement production accounts for about 7 percent of global GHG emissions,[88] it is estimated that it eventually absorbs about one-half of the carbon it originally released in production.[89] New technology allows carbon dioxide to be recaptured in concrete structures. The approach involves injecting carbon dioxide into the concrete mixture while still liquid and as it is being dispersed. The carbon is then captured in the concrete structure, essentially forever. With this procedure the carbon is never released into the atmosphere. In some respects, this approach is similar to that of the CCS.

Note that there are many cases that employ unusual techniques to capture carbon. Typically, however, these involve additional costs. Carbon captured in forest biomass may require a payment to justify the growing and management costs of a carbon-focused forest. CCS requires additional expenditures that would need to be justified either by regulation or compensation for the carbon captured. Concrete capture adds additional costs to the construction process, and so would also require regulation or compensation. Although these are specialized approaches, they all offer the potential to contribute to the overall goal of mitigation.

SOME FINAL THOUGHTS

This chapter has laid out some aspects of an adaptive approach to addressing global warming. My discussion has focused on what are likely to be some of the most disruptive changes. These include sea-level rise, agricultural productivity declines, forest and ecosystem disturbances, and major climate disturbances such as storms and hurricanes. Adaptation (Plan B) is accepted as a supplement to the ongoing mitigation approach (Plan A). Rather than trying to intercept the emission of warmth generating GHGs, Plan B could operate in two different ways. One would be to accept the release of GHGs into the atmosphere, even while trying to control or limit them, but at the same time focus intensively on anticipating, containing, and minimizing the damage created though the application of adaptive management. The second approach would involve the application of one or several of a variety of geoengineering approaches. These involve neutralizing the warming gases once they have entered or are about to enter the atmosphere. This approach would include increased research to understand and develop the potential of geo-engineering types of climate adaptation approaches. I believe the geoengineering has great potential both in itself to contain warming, and also as a supplement to mitigation as it is being undertaken in Plan A. Many geoengineering approaches appear to be relatively inexpensive,[90] especially as compared to mitigation, and can be viewed as constituting a backup safety net against eventual climate change. However, currently the funding for research in this area is very modest.

CONCLUSIONS AND SUMMARY

I have noted above a variety of likely damages associated with warming and have suggested adaptive approaches to address these. I note also that the adaptation approach is not yet well-developed.

More work is needed. Many adaptive efforts will emerge as we anticipate climate changes and their likely disturbances. Flexibility in
response capability is critical to adaptive management. The experience of Puerto Rico in response to Hurricane Maria in 2017 is
an example of the problems associated with a major natural disturbance, together with a lack of appropriate preparedness.

Much of the damage that is likely to occur is simply an extension
of the devastation we have experienced in earlier times from previous
extreme weather events. Hence, we have already developed some of
the tools and management skills that will be needed to address future
climate damage. The challenge today is to reallocate a larger portion
of our resources and efforts to anticipating future damage and developing more effective approaches to addressing these types of events.

Although mitigation may proceed with vigor, it appears unlikely
to be able to limit the human-generated global emissions of carbon
and GHGs to the point where it diminishes the damage due to
warming. This is not only because of the failure to include the effects
of natural warming but also because of the difficulties in smoothly
replacing fossil-fuel energy systems with those using renewables.
Although this transition may be possible for some countries, it is
unlikely to be accomplished in a timely manner globally, particularly
if the developing world is experiencing substantial economic growth.
Thus, according to most climate model projections, much higher
levels of atmospheric GHGs appear likely. With a purely mitigation
approach, such as Plan A, the reduction of GHGs is unlikely to be
adequate to turn the tide on a global scale. Clearly, a Plan B, with its
focus on adaptation, is needed both to complement our mitigation
efforts, and as a backup insurance policy.

CHAPTER 5

ADAPTATION THROUGH REFLECTIVITY AND GEOENGINEERING

I n 1991, Mount Pinatubo in the Philippines erupted, resulting in a massive emission of expanding gases and particulate aerosols. This was followed by a worldwide decline in temperature. Expanding gases cause cooling when the molecules slow down as the gases expand.[1] It is estimated that the eruption caused a massive release of sulfur dioxide that decreased the earth's temperature by about a half degree Celsius. Although the temperature decline only lasted a few years, the influence of the change on the composition of Earth's atmosphere as it relates to climate was clear.

In 1815, Mount Tambora in Indonesia experienced one of the strongest known eruptions. Historical records estimate that global temperatures were reduced an estimated average of 2 degrees Celsius around the world. Europeans called 1816 the "year without a summer," and agriculture production fell for the next several years. This longer term effect is believed to be due to the persistence of volcanic particles that intercept the sun's energy before it reaches Earth.

It is now well-known that certain pollutants can increase or reduce the atmosphere's energy-capturing ability. Examples include carbon dioxide, methane, nitrous oxide and other GHGs that promote global warming. From nature we observe that volcanic eruptions have released gases and particles that have caused global cooling. Other human sources such as industrial sulfur dioxide emissions, which are similar to some volcanic gases, are also known to result in a cooler atmosphere. Sulfur dioxide promotes an increase of sulfate, an aerosol, that prevents the atmosphere from absorbing sunlight and thus results in atmospheric cooling. Sulfur dioxide has health risks for humans, which was part of the justification for the 1990 Clean Air Act. A legislative reduction in certain industrial pollutions, including sulfur dioxide, as a result of the Clean Air Act[2] appears to have had the net effect of increasing temperatures in the United States. Apparently the cooling effects of the diminished sulfur dioxide were offset by the other pollutants that were temperature neutral or, as with carbon dioxide, caused some warming.

Climate engineering, a type of geoengineering sometimes called solar radiation management, involves a deliberate large-scale intervention in Earth's climate system with the aim of mitigating the adverse effects of global warming.[3] Climate engineering is concerned with developing techniques that can modify the atmosphere in such a way as to offset the effects of GHG warming and so specifically affect warming due to the accumulation of carbon dioxide. One approach builds off of the notion that humans can create interventions that mimic the temperature effects of volcanoes. In so doing, humans could affect Earth's reflectivity, the ability of the planet's atmosphere to prevent the warmth of the sun's rays from being absorbed. Geoengineering has been hailed in some circles as a potential technological fix to the climate change crisis. Geoengineering, however, is also vilified in other circles as being risky.[4]

The mitigation approach tries to control warming by attempting to prevent it through dramatic reductions in GHG emissions into the atmosphere that enhance the sun's warming effects. As I have pointed out, adaptation offers an alternative that may be used in concert with mitigation or as an independent approach. One aspect of such adaptation can be found in geoengineering. Instead of trying to prevent the GHGs from building up in the atmosphere, perhaps their effect could be neutralized upon or after release of the gases. In some sense, this approach is similar to dealing with an infestation not by trying to kill all the pests directly, but by neutralizing them, perhaps by sterilization that will prevent their procreation and promote their longer term die-off.

The concept of most geoengineering approaches is to offset or neutralize the warming effects of carbon dioxide. Although there are, as we have discussed, natural mechanisms for cleaning the atmosphere of GHGs, they may take many years, if not millennia,[5] and be of limited use for reducing current warming. However, human activities can assist by contributing to the cleansing of carbon dioxide from the atmosphere using geoengineering measures. These include land management[6] that promotes vegetative growth to capture atmospheric GHGs, bio-energy with Carbon Capture and Storage (CCS), and Direct Air Capture (DAC). All of these involve the prevention of carbon dioxide from entering the atmosphere or the direct removal of carbon dioxide from the atmosphere via biological, mechanical, and chemical means.

Many researchers maintain that removal of existing carbon dioxide from the global atmosphere is required to prevent the earth from succumbing to a long-term warmer climate.[7] However, most of the current efforts are directed not at the reduction of existing carbon stock, but at mitigating current and additional future emissions.

Led by the UN and the current Paris Climate Agreement, the existing mitigation approach pits more than 195 countries against the climate threat. The basic approach is to "slam on the GHG emission brakes" by dramatically reducing the use of fossil fuels. To be successful, coordination and collaboration will be required. The challenges are not only technical but also social, economic, and cultural, due to the international coordination and cooperation that will be needed to succeed in these areas. Most major economies will need to be involved to a substantial degree, but will all nations cooperate fully? Probably not. Can sufficient effort be brought to bear to limit the warming? That remains to be seen. So far, according to the assessment of this book, the future does not appear to be overly sanguine.

As discussed in Chapter 4, adaptation is already underway in an informal sense, as countries and regions around the world try to deal with their localized warming-related problems. An advantage of adaptation is that a high degree of international coordination is not needed. Indeed, often it may not be desirable.

GEOENGINEERING AS ADAPTATION

Earlier discussions of adaptation in this volume referred largely to anticipating and addressing the damage caused by climate warming. Geoengineering is a different form of adaptation that is often unique to a particular area or situation. In many cases, geoengineering need not require wide cooperation. Unlike the local type of adaptation discussed earlier, the benefits of geoengineering are often not confined to one location. An interesting and potentially important application of geoengineering, for example, can involve activities touched on earlier that impact the reflectivity of the atmosphere. One country can undertake these types of activities alone or in cooperation with a few major countries. They don't in themselves require wide international cooperation, and the benefits affect the earth broadly, unconfined to one location, region, or country.

A separate but similar type of geoengineering application may involve the reflectivity of the earth's surface directly, called *albedo* (from the Latin word for "whiteness"). With more reflectivity, the earth's atmosphere will capture less heat. In this case, broad geographical applications will need to be involved to manipulate the planet's landscape and land surfaces to increase their reflectivity. The effects of this climate-influencing approach, however, will not be so much local as global. For this approach to be effective in influencing global climate, it is likely to require large areas, probably involving cross-nation cooperation of a few large countries in order to have a significant climatic impact.

Other adaptive activities are also available and can be utilized to capture and/or neutralize carbon dioxide after it has been emitted into the atmosphere. This general approach, which resembles mitigation activities, can be undertaken by an individual country or group of countries. Such activities may have little global climatic effects unless they are widely practiced and combined with a broader mitigation-like program.

An important concern with geoengineering activities relates to our ability to intervene in natural systems without causing harm. Do humans know enough to intercede efficiently and safely in a major natural process such as the atmosphere? Are the potential risks of the application of new warming-preventing technologies too great? Opposition to atmospheric geoengineering can already be found[8] within the environmental community. One notable concern relates to the potential effects on agriculture of increasing reflectivity. While increased reflectivity will decrease the earth's heat absorption, the reduction in sunlight could concurrently reduce the productivity of agricultural crops.[9]

Despite the risks and uncertainties of geoengineering approaches emphasized by some, others argue that humanity appears to have little or no choice.[10] We inadvertently opened modern technology's Pandora's Box in the massive use of fossil fuels that had the unintended consequence of changing our atmosphere and our climate. We have created global warming by virtue of the needs of our high-energy economies. Now we must be prepared to confront the consequences, either with or without the use of further technological innovations.[11]

TYPES OF TECHNICAL INTERVENTIONS

Let us examine some potential geoengineering approaches to achieve adaptation, approaches that have the potential to take GHGs out of the atmosphere or at least offset their global warming impact. As noted, the term geoengineering covers a host of interventions and involves a variety of different climate-managing techniques. Some of these are scientifically sophisticated, but not all. An intended consequence of climate engineering is to provide the capacity to neutralize, offset, or reverse in some manner the influences that cause climate change. One such approach commonly involves interventions that intend to neutralize the effects of the GHGs and thereby stabilize the earth's reflectivity. Specifically, it enhances or maintains the viability of the earth's atmospheric shield so as to offset the warming effects of GHGs, most particularly carbon dioxide. An increase in atmospheric reflectivity can decrease the earth's absorption of the sun's energy and thereby prevent additional warming that the increased GHGs might have generated.

Different methods are being considered. One proposed idea involves placing reflective particles in the atmosphere, while another suggests using complicated techniques for enhancing the whiteness of clouds, thereby increasing the reflective capacity of the earth's atmospheric shield. Still another proposes placing aerosols or chemicals with properties similar to those of sulfur dioxide in the atmosphere to reduce its ability to capture energy, as has already been seen with the cooling effects of certain industrial pollutants. Such approaches generally have a temporary impact and therefore are in principle reversible. If the likely positive effects are to persist, these approaches would require periodic recharges of the atmospheric material.[12] Other reflective techniques may focus on the land surface rather than the atmosphere. A conceptually simple approach focuses on changes the earth surface's, reflectivity[13] which can be accomplished by replacing a high heat absorptive ground cover with a more heat reflective one. For example, a reflective grassland would replace a heat absorbing forest.[14] The more energy reflected, the less warming will result.

In this chapter, I examine a set of control approaches in which humans adapt to warming not by mitigating or preventing a basic cause but rather by adapting to the effects of GHG emissions with actions that neutralize or offset the warming process. Thus, most of the approaches discussed in this section do not involve preventing the use of fossil fuels. Rather, the technologies allow emissions to occur but then permit us to undertake adaptive management interventions that show some promise of countering any warming effects that may occur. In short, such interventions prevent the GHG emissions from promoting warming by neutralizing the warming effects of the gases or, in some manner, capturing and/or disposing of the emitted gases. Note that geoengineering interventions rarely are in conflict with ordinary efforts to reduce or eliminate emissions, but they can be undertaken independent of or complimentary to mitigation approaches.

SOME INTERESTING APPROACHES

With so many intervention options available, possibilities run between the ordinary and the bizarre. One of the more unusual proposed geoengineering approaches is to capture already emitted carbon and entomb it in the sea bottom.[15] This approach involves promoting the productivity of krill, a very small crustacean that is found in the oceans of Antarctica. Aggregate krill volumes are large, and they are at the base of the food chain for most animal life in Antarctica. Most ideas involve ocean fertilization, which consists of adding nutrients in the form of iron compounds to the ocean. This change would increase primary production of krill substantially, since the krill feed on algae that would thrive on the iron and thereby pull more CO_2 from the atmosphere as it proliferates. This fertilization technique would be aimed at the southern oceans, which have the greatest potential for krill growth. The krill process large amounts of carbon from the ocean vegetation that consumes the fertilizing compound. The waste that the krill create and waste caused by other ocean creatures that eat the krill then functions to funnel the carbon toward the ocean floor where it is to be captured.[16] Krill feed on surface algae but defecate at great depths, thereby acting as a conduit to funnel the carbon-laden feces to long-term captivity in the ocean floor. Ocean fertilization is believed to enhance the total number of krill to the extent that potentially very large volumes of carbon can be captured in this manner. However, some researchers now view krill as potentially threatened by warming waters, which could complicate such adaptive efforts.[17]

Krill are not the only natural systems that can be harnessed for capturing previously released atmospheric carbon. Other approaches include restoring previously existent forest areas (reforestation) and creating entirely new forests where none had existed before (afforestation).[18] The world's forests hold captive huge volumes of carbon in their wood stocks. Biological growth involves the capture of the carbon from the carbon dioxide molecules in the atmosphere and the release of an oxygen molecule back into the atmosphere. This process removes a primary GHG, namely carbon dioxide, from the atmosphere, while releasing an oxygen molecule, O_2, into the atmosphere. Large expansions of forests, with the long lives of the trees they nurture, allow for the long-term capture and holding of carbon. Symmetrically, forest contractions can allow for rapid carbon dioxide release through, for example, wildfires.[19] Global forests are large enough that significant changes in either direction can have a substantial effect on overall global atmospheric GHG levels. Various systems, often ones involving carbon credits for net carbon capture by new forests, have been considered to promote the growing of new forests primarily for carbon sequestration purposes.[20]

GEOENGINEERING AND THE IPCC?

So far the Intergovernmental Panel on Climate Change (IPCC) has given relatively little attention to geoengineering options. In the entire Third Assessment Report, only a small part of one chapter on biological sequestration was devoted to geoengineering.[21] As co-chairs[22] of that chapter, my colleague and I were both well aware of how much this topic was being short-changed since our authors' group had only limited expertise in this area. This was also particularly true in the geo-engineering area of solar radiation management, which to my mind seems to have substantial potential but was ignored in our chapter.

GEOENGINEERING RESEARCH

Few of the climate engineering techniques discussed have been fully developed. Most are in their research stages. Funds for research in many geoengineering areas have been limited. Recently, as reported via the Internet, the Congressional Committee on Science, Space, and Technology, chaired by US Representative Lamar Smith (since retired), had been discussing geoengineering. "As the climate continues to change, geoengineering could become a tool to curb resulting impacts," said Smith. "Instead of forcing unworkable and costly government mandates on the American people (such as GHG mitigation), we should look to technology and innovation to lead the way to address climate change. Geoengineering should be considered when discussing technological advances to protect the environment."[23]

Rep. Smith, sometimes identified as a climate denier, was especially critical of the progress of those with a traditional GHG climate view. "Climate alarmists have failed to explain the lack of global warming over the past 15 years," Smith said in 2016.[24] "They simply keep adjusting their malfunctioning climate models to push the supposedly looming disaster further into the future." Smith, however, was less skeptical about climate change per se when saying at a congressional hearing that "climate is changing and humans play a role" and suggesting that it's now just a question of the "extent" to which human activity is the climate-change driver. So perhaps geoengineering, often characterized as a false solution to the climate crisis,[25] may become the savior of sorts, as it appears to offer potential as an important backstop to fossil fuel mitigation. This issue is discussed below.

THE RATIONALE OF GEOENGINEERING

As discussed above, an alternative approach to preventing GHGs' warming effects would involve a program to modify the atmosphere in a manner so that it would neutralize the effects of these gases by offsetting their ability to capture solar energy. As I have noted, there are a number of adaptive technological approaches can address climate change. In general, they influence the atmosphere after GHGs have been released but before they can contribute to warming.

A general type of geoengineering approach that might mimic a cooling effect could involve the injection of a sulfate aerosol, such as sulfur dioxide (SO_2), into the atmosphere on a periodic basis. Since a kilogram of sulfur would offset the negative impact of several hundred thousand kilograms of carbon dioxide, the result would be a reduced amount of the sun's energy captured.[26] Such a phenomenon is similar in many respects to the eruption of a volcano where large amounts of particles are created. These materials, which modify the atmosphere, tend on balance to have the effect of offsetting the warming effects of GHGs. As I shall discuss below, sulfur dioxide, which is associated with human health problems, is not the only gas that may be useful as a global coolant.

An issue with many geoengineering approaches, and especially with the reflectivity-type approaches, is what we have come to know as the "free rider" nature of the climate solution.[27] Note that in these situations all parties have the incentive to wait and let someone else provide the solution. Thereby, the development and application of a comprehensive solution could result in most of the world receiving a "free ride," with the burden of development and implementation falling disproportionally on one or more of the lead countries. On the other hand, if too many countries tried to initiate independent activity simultaneously, excess cooling might inadvertently be generated. Thus, although a few countries could, in principle, undertake critical activities, coordination would be essential.

SOME CONTEMPORARY GEOENGINEERING PROJECTS

Although I noted the absence of funding and research into geo-engineering applications, all is not bleak. Harvard University has been undertaking experimentation with the object of determining whether a geoengineering technology can safely simulate the atmospheric cooling effects of a volcanic eruption.[28] If successful, this approach could provide an additional adaptation tool to address climate warming, should it be needed. The promising approach would involve modifying the atmosphere with calcium carbonate particles to offset the warming effects of increasing GHG levels. This type of approach would involve mimicking the effects of a volcanic eruption. As noted, the 1991 Mount Pinatubo eruption lowered global temperatures for multiple years, and evidence suggests that some earlier volcanic eruptions like Tambora have had worldwide cooling effects that persisted for years. A program designed to provide periodic injections of particles, as Harvard's study suggests, could be managed to create a continuing cooling effect. The Harvard scientists' approach involves sending aerosol injections sixty-five thousand feet into the earth's stratosphere. This is part of one of the world's biggest solar geoengineering experiments to date and designed to study a potential technological solution for global warming.

The University of Washington had an ongoing experiment to seed clouds with salt water aerosols to brighten them, to see if they will reflect more heat back into space. Climate scientists recognize that clouds are one of the least well understood climate factors. Although they are obviously important, clouds are poorly represented in climate models. In concept, clouds could either promote global warming or promote cooling. The Washington experimental test involves two small-scale dispersals: water followed by the release of calcium carbonate particles, which are hoped to brighten the clouds and promote the increased reflectivity of solar energy.[29] Future tests could involve seeding the sky with other types of particles to see how well they intercept the sun's energy.

Although not the only studies of climate geoengineering, these are among the largest and the most comprehensive to date. It is estimated that the development of a geoengineering solar buffer could protect the earth for as little as $10 billion annually.[30] Note that this amount is only one-tenth of the annual sum projected in the Paris Agreement to be distributed to the poorer countries to aid in their global warming efforts. Individual countries are also spending huge sums to restructure their energy systems. Geoengineering is potentially fast and cheap and may prove to be a powerful adaptive tool to supplement other approaches or provide a critical backup should the Plan A approach fail or fall short of expectations.

Other somewhat different geoengineering approaches are under development elsewhere. An example is a gas-guzzling bacteria under development in New Zealand. This research involves the use of a type of methane-consuming bacteria found in some of the world's most extreme volcanic environments. The bacteria gas oxidizers could be used to mitigate methane and other greenhouse gas emissions, according to a study published in the *International Society for Microbial Ecology* journal.[31] The bacteria are already used in industrial settings to turn methane emissions into fuels and protein feeds, but scientists hope they can have broader applications.

Finally, the National Center for Atmospheric Research (NCAR) has been using model projections to ascertain the implications of different "shading" approaches to reduce the heat capture from the sun.[32] The study is based on the injection points of the aerosols' introduction into the atmosphere. (Alternative release points, at different latitudes, show different long-term climate effects.) Generally, these reduced negative side effects in the weather that seemed to occur with earlier injection placements. The research is still underway, and the full complement of climate and weather effects is not understood. An additional rationale for this approach is that the mitigating effects of attempts to reduce GHGs have been slow to be realized. This solar radiation management approach is viewed as having the potential for generating a much faster response to controlling warming.

The Potsdam Institute for Advanced Stability Studies has been looking at the potential side effects of "shading," particularly the effects of a sudden stoppage of the injections. They found that there would be a lengthy buffer effect that could allow for a follow-on corrective response, so that negative effects of sudden stoppage could be minimized. The Potsdam group also estimated the costs of such an approach at $50 billion initially and $12.5 billion annually to operate.[33] Note that this cost is substantially higher than some earlier estimates.

One of the most difficult aspects of atmospheric climate change modification is that of influencing clouding. This is a critical part of these three processes. Climate scientists acknowledge that clouds are important to climate change and warming, yet their role is neither well understood nor well represented in climate models. It appears that some types of clouds promote warming while others have the reverse effect. For example, stratus clouds are believed to have a net cooling effect while cirrus clouds are believed to promote warming.[34] Influencing the nature of clouding is believed to be a major key to obtaining the potential for managing climate change.

The idea behind cloud brightening, for example, is that seeding marine stratocumulus clouds with sub-micrometer sea water particles may significantly enhance the droplet concentration of the cloud and thereby increase its reflective capability. This could produce a level of cooling that would have the ability to balance global warming at some acceptable point. This approach envisions distributing water into the atmosphere using either mobile or stationary sources. One proposal suggests a fleet of sea-going vessels devoted to this effort. Research is currently being undertaken on all of these possible warmth-controlling technologies.[35]

Despite gaps in our knowledge, cloud projects involving the use of marine brightening have been underway in addition to stratospheric aerosol particle injections and solar radiation management involving the direct injection of particles into the stratosphere, as in the Harvard and University of Washington studies discussed above. The injected particles scatter solar radiation back to space and reduce the solar energy captured by the earth. The early investigations focused on sulfuric acid particles, which mimic volcanic injections of an aerosol. However, a wide array of particles and chemical compounds appear to have a much greater potential than sulfur aerosols for scattering solar radiation. Questions arise as to the consequences of such particle injection, particularly with regard to their consequences for the ozone layer. There are questions as to how the injections might chemically react to the ozone and if this might cause problems. There is a question of acid rain that might be associated with injected chemicals. Indeed, much of the ongoing research is examining these types of issues.

CARBON CAPTURE AND STORAGE (CCS) AND OTHER ADAPTIVE APPROACHES

As was briefly mentioned toward the end of the previous chapter, there are a variety of mechanical engineering approaches and natural, biologically driven approaches to address GHGs about to enter the atmosphere or already in the atmosphere. One of these would involve ambient air capture that could simply involve a mechanism that would capture carbon dioxide from the air. Currently, the most common carbon dioxide capturing devices are associated with Carbon Capture and Storage (CCS). This system allows for the capture of carbon from the flow of carbon dioxide gases in the exhaust gas streams released from fossil-fuel energy power plants. This process can eliminate 90 percent of the carbon dioxide released from the energy process.

The process involves a technique similar to that commonly used to remove pollutions from their gaseous waste streams. The method involves bringing the gas waste stream in contact with a fluid that removes gaseous components into the fluid. In essence, the fluid absorbs the undesired polluting gases. This approach is similar to current methods for controlling air polluting gases, such as sulfur dioxide, from industrial facilities. For capturing carbon the technique is similar and involves "scrubbing" the carbon dioxide gas out of the gaseous waste stream as it is being released following the process of burning a fossil fuel. These approaches involve separating the pollutants from other waste gases in the production process, essentially while on the way to the smokestack. The focus of the process is on capturing carbon dioxide before it is released into the atmosphere. Once captured, the issue becomes how to dispose of the captive carbon.

The anticipated approach to disposal has been to inject the captive carbon dioxide into the geological caverns in the earth. After the carbon dioxide is captured, it is compressed into a liquid and transported elsewhere to then be injected into porous rock and, if properly located, is trapped by the geology and prevented from escaping. Over longer periods, much of the carbon can be captured as minerals in porous rock, old oil formations, and other geological formations. In fact, carbon dioxide has been transported and effectively injected into geological formations for decades to promote enhanced oil recovery by increasing the pressure on the oil within rock formations to make its extraction easier. The now liquid carbon dioxide can also be transported via pipelines to distant locations.

Although still in their developmental stages, approaches of this type for the capture and transport of carbon dioxide are underway in prototype applications by the Norwegians in the North Sea oil fields, by the Canadians in their fracking operations, in Australian shale fields, and also in the US shale oil fields. Statoil, the Norwegian oil company, has undertaken a host of applied CCS studies in its North Sea oil facilities,[36] and it is considering making this approach broadly operational. The California Energy Commission is considering the greater use of CCS to address some of its carbon dioxide emission problems.

In the United States, additional research is in progress. In North Dakota, two exploratory wells are under development in the middle of coal country to help researchers determine the feasibility of storing coal-generated carbon dioxide underground rather than emitting it into the atmosphere. Researchers are investigating the geology more than a mile underground to determine if it is suitable for the storage of carbon dioxide captured from nearby coal-based energy facilities. For example, a study in New Brunswick involves the injection of carbon dioxide with the objective not only of testing the behavior of the carbon dioxide; it will provide researchers with more information about local geology.[37]

The Department of Energy study is looking to determine the feasibility of injecting two million tons of carbon dioxide per year for twenty-five years. The researchers' ultimate goal is to use carbon dioxide captured from coal-based facilities for enhanced oil recovery, allowing oil producers to extract even more oil from existing wells. Recently, North Dakota, Montana, Wyoming, and the Canadian province of Saskatchewan signed a memorandum of understanding to share research on carbon capture utilization and storage. Additionally, the US has recently modified its tax code to expand existing tax credits for CCS capture, storage, and utilization projects.[38]

CCS brings a number of advantages and disadvantages.[39] First, it allows for the continued use of fossil fuels for creating certain types of energy while eliminating their negative climate effects. Second, underground storage of carbon dioxide could be a good backup plan should the development of renewable fuels be delayed. A major problem with the CCS approach is its high cost, often doubling the cost of coal energy.[40] In the long-term, CCS appears unlikely to be cost-competitive with renewables in many applications. In addition, there are groups that oppose the construction of new coal-powered plants. The argument is that additional coal plants will bring about pressure to utilize coal, whether or not CCS turns out to be effective. Furthermore, renewable energy such as wind is inherently cheaper than coal use augmented with CCS, and therefore resources ought not to be wasted on CCS.

A final argument in favor of CCS relates to its use with biomass to reduce overall atmospheric carbon dioxide. Biomass can capture atmospheric carbon and CCS can store it permanently in geological formations. So, attempts to reduce overall atmospheric carbon levels, which might become important in the future if we are not successful in controlling GHGs now, could use the application of CCS on biomass-fueled power operations. In that system, biomass would capture carbon and provide energy, while at the end of the cycle CCS would sequester the gas permanently in geological formations. The net result would be an action that reduces carbon dioxide in the atmosphere while producing energy for humans.[41]

Soils, too, can be managed to increase or decrease their carbon stock. For example, the carbon content of soils can be enhanced by appropriate management of forest ground cover and the plants that compose it, and the carbon in agricultural soils can be increased through management such as "no-till" cropping techniques, which protect the carbon in the soil from the atmosphere and its oxygen.

Trees and soils are sometimes referred to as a carbon "sink" with their ability to capture large volumes of carbon.[42] They can also be a "source" by releasing carbon. Land management approaches can move terrestrial systems from sink to source and back. Although the ability of the terrestrial system to capture carbon has limits and is not sufficient on its own to prevent global warming, it can be large enough to accelerate or reverse the buildup of GHGs.

Some have argued that terrestrial system management can "buy time" to allow for the development of more permanent systems to mitigate GHG buildup, and therefore help meet broad climate goals. [43] Discussions have been underway for some time to identify approaches to provide incentives to capture carbon. These systems could be financial or in the form of carbon offset credits used as subsidies to tree growers for producing additional wood biomass, with its associated captive carbon.

Using vegetative approaches to carbon management recognizes the inherent dynamism of biological systems. The forest can continue to grow; trees may be felled and burned as well as planted. But both of these objectives can be realized and still allow other climate objectives to be achieved, such as allowing the net stock of biomass to expand over time. Through forest expansion, the stock of carbon continues to be removed from the atmosphere and diverted into the terrestrial system.

As suggested, biomass is a renewable fuel that can be burned for energy. This feature opens an additional option to capturing carbon, but adds a complication for addressing GHGs, particularly carbon dioxide. By burning biomass, the carbon capturing function of that part of the forest ceases. If biomass merely substitutes for coal, little net change would occur. However, if the wood burnt for energy is eventually replaced by new growth, the overall process can be carbon neutral in the longer term, and bioenergy from this source can be viewed as a renewable.

Analysis has pointed out that creating new forests on a massive plantation basis could capture enough carbon to have a significant effect on global GHG emissions. They note, however, that the process would not be costless and could involve large volumes of land, water, and nutrients.[44] Activities of this type are ongoing. Many European countries use substantial volumes of wood, largely in the form of wood pellets, to assist in meeting their GHG emission reduction targets. Large portions of these pellets originate from planted forests in the United States.

Finally, there are other approaches involving bioenergy and the use of biomass material. Other materials, such as biochar, involve the creation of a charred form of biomass. The decomposition of biochar is very slow, much slower than the decomposition of ordinary logs. It is sometimes suggested that burying biochar will lock up the carbon held captive for very long periods of time.[45]

ALBEDO ON EARTH'S SURFACE

Land management functions as another geoengineering approach to manage the reflectivity of the earth's surface. This approach is separate from efforts earlier discussed to enhance the reflectivity of the planet's atmosphere. The amount of energy reflected by Earth's physical surface is referred to as the planet's albedo.[46] By modifying the nature of the earth's surface, the reflectivity can change, and greater reflectivity means less solar warming. Desert sands have, for example, an albedo level of about 0.40, meaning 40 percent of the sun's energy is reflected. By contrast, forests have an albedo level between 0.08 and 0.15 of the sun's energy input, while fresh snow has an albedo level of between 0.88 and 0.90.[47] Thus, forests of the far north that maintain snow on their canopies for extended periods throughout the year are a useful reflector of the sun's energy and a factor in reducing global warming.

Although the average global albedo is estimated at about 30 percent, variation creates potential management opportunities. Note that there may be a conflict between the global cooling effects of capturing GHGs in a forest and the warming effects of the lower albedo associated with forests or other land cover. An example may be found in the trade-off between grasslands and forest lands. Although forests have desirable properties in capturing warmth-creating GHGs, they also tend to have a low level of reflectivity. Therefore, clearing a forest has a duel but contradictory effect on warming. The release of GHGs promotes warming, but the conversion of forest lands into grass increases reflectivity with its associated cooling effect.[48] So, Mother Nature being complicated, those who are cutting down the Amazon rainforest could be seen by some as countering global warming instead of aggravating.

Additional research addressing the earth's reflectivity is underway in the Arctic. Glaciologists are examining the possibilities and implications of covering part of Greenland's ice sheet with reflective materials to further slow the melting process.

This is not a long-term solution, but by slowing the melting process, time can be bought in order to develop more effective basic approaches to deal with climate warming. Slowing the melting process in itself can provide a supplement to natural surface albedo that will retard the growth rate of temperature. Other approaches under discussion include those to prevent warm water, for example from the Gulf Stream, from melting glaciers.[49]

Thus, land management must be applied in a sophisticated manner, with considerations of both the benefits and disadvantages of altering the landscape. Large areas of land must be involved in order to have a significant effect. Some conflict between usual desired land uses—cropping, pasture, forestry, development—and global warming objectives appear highly likely, as things like pasture may be more profitable ventures, but forestry more climate-protecting.

AN ADDITIONAL ISSUE: PULLING CO_2 OUT OF THE ATMOSPHERE

Even if new GHG emissions were to cease flowing into the atmosphere, the cumulative GHG stock buildup would remain large well into the future and would keep global temperatures at high levels. It is estimated that between 65 and 80 percent of carbon dioxide dissolves into the ocean over a period of twenty to two hundred years.[50] Today there is rising interest in negative emissions technologies that pull carbon dioxide from the atmosphere as a way to bring temperatures downs if they surpass acceptable levels.[51] Additionally, if undertaken soon, negative emissions strategies could help nations keep global temperatures from exceeding the targeted 2 degree Celsius of the Paris Agreement.[52] Negative emissions technology offers the hope that climate goals can still be met, even if temperatures initially rise above 2 degrees Celsius, since removing carbon dioxide from the atmosphere should cool the planet.

Negative emissions, although possible with natural systems such as forestry, would generally require some form of geoengineering. Researchers have proposed an array of negative emissions strategies, involving everything from futuristic machines that absorb carbon to common carbon-sucking tree plantations. Although now largely conceptual, current research is looking into the potential feasibility of innovative approaches. Carbon sucking machines are technically feasible but appear likely to be expensive and of limited use.

The upside is that the techniques involved in tree planting, planted forest management, and no-till cropping are well understood and relatively cheap. However, the downside for forests is that they require a lot of space and preclude other land use activities. In addition to the trees, forest soil is effective at storing carbon when it's managed properly. No-till agriculture,[53] however, could be practiced on lands that are already producing agricultural products, and therefor produce another critical service in the form of carbon storage.

There are other approaches to promote landscapes that could be soaking up even more carbon than normal. Biochar—wood or agricultural waste that has been turned into a charcoal-like substance through exposure to heat—can be a vehicle for absorbing GHGs from the atmosphere. The vegetation captures carbon dioxide, and when turned to biochar, its decomposition and carbon release are severely retarded. Thus, carbon dioxide emissions are inhibited. Biochar can also boost the productivity of cropland, helping soil retain more water and nutrients and soak up more carbon.

Crushing rocks offers another method to capture carbon, albeit a very slow one. As discussed by David Archer, the global carbon cycle involves cycling through a variety of forms including rock materials.[54] The geochemical carbon cycle involves millions of years.[55] Crushed rock can speed up the weathering process and is sometimes known as enhanced weathering. When rain forms, it dissolves a little bit of carbon dioxide from the atmosphere. The slightly acidic rain wears away at rocks when it falls to the earth, producing a chemical byproduct called bicarbonate, which mixes into the soil. Some research indicates that enhanced weathering could put away several billion tons of carbon dioxide each year if applied at a global scale. David Archer's[56] discussion of the long-term carbon cycle deals with carbon's flow in and out of mineral rock.

One of the most widely discussed negative emissions proposals involves not just planting forests, but also harvesting them for energy.

It's known as bioenergy with carbon capture and storage, or BECCS. The idea calls for large tree plantations to draw carbon out of the atmosphere into the trees. The trees are then cut down and used as a biofuel substitute for fossil fuels. A carbon capture technology, much like CCS for coal-powered plants, would collect and store the resulting emissions from a wood-fired power plant before they could escape back into the atmosphere. The net effect would be more than rendering the entire process carbon neutral. The approach would, in effect, transfer carbon from the atmosphere to a permanent geological storage facility.[57]

CARBON STORAGE

A nagging issue of some of this technology is that of the storage of the captured carbon. The most commonly suggested solution is to pump the carbon dioxide back into below-ground geological formations. As discussed elsewhere in this book, most operating facilities are using this approach.

There has been a bit of a search to find useful applications for the captive carbon dioxide both to provide a disposal point for the CO_2 and to provide additional income to finance the activity. As noted, CO_2 can be used as a pressuring method to force more oil from existing oil formations. Other fledgling applications include its use as a component in concrete production (see Chapter 4). The concept of a lithium/carbon dioxide battery that would utilize CO_2 is attractive, although the volumes used are likely to be small compared to the volumes that potentially could be captured.

The Bioenergy Carbon Capture Storage (BECCS) strategy is attractive in that it promises both carbon removal with plant growth, and clean energy production while substituting a renewable plant for a fossil fuel. But major concerns about its environmental impact on a large scale have arisen, mainly related to the sheer amount of water, land, and other natural resources that would be needed for massive tree plantations. However, if plantations are placed on areas that receive adequate natural precipitation, water need not be a problem. Studies vary widely in their estimates of how much carbon BECCS could be stored globally.[58] The literature review released in 2018 shows that the range may run anywhere from 0.5 billion to 5 billion tons of carbon dioxide annually by midcentury.[59] There is still the question of whether the forests might be placed in areas more suited for crop production.

Some experts propose using machines, rather than natural carbon sinks, to draw carbon dioxide out of the atmosphere. Last year, the Swiss company Climeworks became the first to open a commercial carbon-capture plant, currently operating outside Zurich. It uses special filters to capture carbon with a chemical bond, making it easy to collect.[60] While direct-air capture is now proven in small projects, its potential at a large scale remains to be seen. The technology's upfront costs are among the greatest barriers to its advancement. But there are other concerns as well, namely the amount of energy required to drive the machines. Unless the electricity comes from carbon-neutral sources, there's the risk of negating some of the benefits of the carbon dioxide being pulled out of the air. Another concern is that of an over-reliance on mechanical carbon capture, a technology that has yet to be proven, that could distract from more fundamental mitigation.[61] In other words, there might be an increase in CO_2 release because many believe that we can just easily wash it out of the atmosphere, thus defeating the purpose of the technology.

Studies on individual negative emissions strategies, such as BECCS or direct air capture, suggest that even when applied at a global scale—accounting for factors like the amount of land, power, or other resources required to maintain them—they might each store a few billion tons of carbon dioxide each year, on the high end. If multiple technologies were deployed together, though, their combined carbon-capturing potential could be far more substantial.

OTHER CONCERNS WITH GEOENGINEERING

Geoengineering approaches are not without their critics, however. The types of activities envisaged as geoengineering are so different that concerns appropriate to one set may be meaningless for another. The general concern is whether any of the geoengineering approaches can be benign in their application. Can we avoid technologies that will be seriously disruptive to important and significant ecosystems? Can the types of atmosphere-altering systems under examination avoid being permanently disruptive to the atmospheric system? Can they avoid disrupting ocean systems, land and forest systems, and so forth? These are questions researchers obviously should examine.

It is apparent that some technologies can cause problems if not applied carefully. For example, we know that emissions of sulfur dioxide from industrial pollution or gaseous emissions from some volcanoes can reduce warming, but many have negative effects, like sulfur dioxide health risks for humans. Simply offsetting the warming effect of increased carbon dioxide many not neutralize side effects. Although cutting incoming solar radiation via atmospheric manipulation could reduce warming, it has been hypothesized that it may also affect the hydrological cycle in ways that could promote drought. Another concern already anticipated for GHG and carbon dioxide buildup in the atmosphere is that of the impact of high atmosphere carbon on the ocean's acidity level. For example, acidity of the oceans may continue to rise even if warming were halted but emissions continue. That could generate damage to sea life, coral reefs, etc. Similarly, utilizing land for carbon sequestering forests might put pressure on other land uses and resource availabilities, including conflicts with agriculture and water demands.

The general set of concerns with geoengineering relate to possible unintended environmental consequences that may ensue, including the potential for disasters. In the absence of significant moderation of GHG emissions, the forecasted anticipated pace of warming would be the highest in human experience. Although civilized humans have previously experienced warming in this interglacial period, we have never experienced the magnitude of warming anticipated in many current temperature forecasts. We have few certainties about any unintended consequences that might accompany the warming. Thus, whether we experience rapid warming or the initiation of adaptive geoengineering measures that can generate possible unintended consequences, we continue to move into uncharted territory.

SOME NON-TECHNICAL CONCERNS

A major background concern throughout our discussion of GHG mitigation has been that of achieving widespread cooperation among the more than 195 signatories of the Paris Agreement and other future agreements. How important a problem is free riding? It is one thing to sign an agreement and quite another to fully participate. And although not every country needs to participate with full gusto, the major ones certainly do. According to climate analyst Scott Barrett of Columbia University,[62] "The focus of climate policy so far has been on reducing the accumulation of GHGs. That approach, however, requires broad international cooperation and being expensive has been hindered by free riding."[63] By contrast, adaptation, as often found in solar engineering, "will bring a brave new world with issues involving the governance of an unprecedented technology."[64] Such an approach does not require the cooperation, or even the consent, of the many nations that signed the Paris Agreement. But who will develop the solar engineering technology and who will initiate its implementation? These are issues that have not yet been encountered let alone addressed. However, these decisions may be required only a short way down the road particularly if Plan A is found to be as weak as many have suggested and if the world is forced to move to an aggressive Plan B for survival.

There is competition among researchers for funding. More funds directed to geoengineering imply less for traditional mitigation types of research, including the development of renewable wind and solar energy sources. Some senior climate scientists view potential geoengineering developments with alarm, while others believe that much of the concern relates to fears of a cash drain from mitigation technologies, such as wind and solar energy, to alternative approaches. On the same day as the House Science Committee hearing, twenty-four researchers delivered a letter expressing concern about the premise of the hearing that a useful geoengineering approach could arise and move forward.

Other concerns are ideological or ethical. Many believe that humans are overexploiting the earth. Climate change, commonly attributed to human activities, is viewed as one of the most intrusive of these forms of exploitation. Recall again Robert Nelson's discussion of environmentalism as a religion from Chapter 1. People appear to be more comfortable undoing what has caused a problem rather than developing a new technology to override the problem created. According to its critics, geoengineering is a distraction from the actual challenge of halting dangerous levels of greenhouse gas emissions and subsequent runaway climate change. Critics also say that the approach is a potential danger in and of itself because its techniques may well re-engineer the atmospheric system, with perhaps unpredictable side effects.

Naomi Klein, author of the book *This Changes Everything: Capitalism vs. The Climate* has been among the more outspoken critics of geoengineering. "Geoengineering offers the tantalizing promise of a climate change fix that would allow us to continue our resource-exhausting way of life, indefinitely,"[65] wrote Klein in the *New York Times*. "And then there is the fear. Every week seems to bring more terrifying climate news, from reports of ice sheets melting ahead of schedule to oceans acidifying far faster than expected."[66] It seems to me that this is a criticism of the evils of warming and its implications, but not a genuine critical look at geoengineering or its possible solutions. Obviously, a strong ideological underpinning is embedded in much of the criticism and these particular comments.

There is no question but that there is a real concern. Do the new technologies offer more potential costs than benefits? If Plan A works, it might have been very expensive, but it still accomplished its objective. But from the broad overview of the analysis of this book, it appears questionable that the current approach of mitigating GHGs will succeed. Where is the backup should Plan A be inadequate? I would surmise that most of us would prefer survival with geoengineering and Plan B rather than to move toward human extinction as Plan A gradually fails. David Victor et al. present a blunt approach in *Foreign Policy*, asserting that the requisite scientific work developing geoengineering approaches needs to be done even if they are never used.[67] The requisite scientific work developing geoengineering approaches needs to be done even if it is never used.

GEOENGINEERING REALITIES

Given society's great successes in developing technical fixes to address a variety of problems, it is remarkable that so little has been done to begin to apply these types of technological approaches to the problems of climate warming. Humans have always been ambiguous about, if not frightened of, change, innovation, and science generally. We have been successful as a species by being basically collectively conservative. We speculate about the future but often it includes Frankenstein characters, aliens, war of the worlds, and so forth. Dangers appear to abound. Is it better to address the danger you think you know, e.g., warming, or the danger you have not anticipated, such as those that some speculate might be associated with geoengineering approaches? In either case, the future will involve uncertainties for humankind. Our ally is technology, but it may also be our enemy, and we have to anticipate and control for such dangers.

CONCLUSIONS

Geoengineering has many facets. These range from fertilizing the southern seas to promoting krill populations, to means of modifying atmospheric gases and/or clouding to the promote the neutralization of the warming effects of carbon dioxide, to efforts to promote land surface modification to increase albedo. There are techniques to remove carbon dioxide from smokestacks using CCS, or from the atmosphere directly with biological sequestration and relocating it into a forest or the earth, where in some cases it had originally been a component of a fossil fuel.

To either embrace geoengineering or to reject it unequivocally ignores the differences and complexities. Current questions about the safety of geoengineering tend to view the technology narrowly. The question of whether human-generated atmospheric cooling effects are similar to that of a volcanic eruption involves only one of a host of possible approaches. However, it has the potential to be a very attractive approach. Let us hope that we can find technology that is both successful and safe. In this case, the approach could provide an additional adaptation tool to address climate warming, should it be needed. Geoengineering may prove to be a powerful adaptive tool in Plan B—a tool that may supplement other approaches and/ or provide a critical backup should Plan A be inadequate. Indeed, as David Victor and his colleagues have suggested in their paper in *Foreign Policy*, the "geoengineering option may be a last resort against global warming."[68]

CHAPTER 6

POLITICAL CHALLENGES

CONTROLLING GLOBAL WARMING

Global climate change appears to be a type of problem almost perfectly suited to be addressed by coordinated mitigation by a strong central authority. Coming out of World War II, the newly created UN was designed primarily to address international issues that extended beyond the borders of any one country. As such, the UN would appear to be the perfect instrument to address a multi-country problem like human-created climate change. And perhaps it is. Today's concern with climate change is following in the footsteps of some earlier multicountry environmental problems that the UN was instrumental in addressing.

One of these earlier problems was "acid rain." This concern originated in Europe where, in the 1970s, substantial areas were experiencing local forest decline and dieback. Dieback is a situation where trees of the forest are struggling to achieve growth. In the acid rain issue, the leaves and needles were turning brown and the trees were dying prematurely. The culprit was believed to be acid rain, which was created by the emissions of industrial operations that, upon mixing with the air, found their way into forests as acidic rain. This was believed to be the source of damages to the forest.[1] Something similar was occurring in a number of forested areas across the US. The damages were dieback and an increased forest mortality, with dead and unhealthy trees sometimes occurring in bunches but often as single victims. The dieback, not easily attributable to infestation or the usual diseases, was initially attributed to acid rain, which was largely generated from sulfur dioxide emissions. Although localized, the widespread occurrence of this problem, particularly in Europe, gave it a sense of being a global problem. Thus it could be viewed as a global externality.

Subsequently, the passage of the Clean Air Act[2] in the US, which imposed restrictions on industrial emissions and especially sulfur dioxide thought to be responsible for acid rain and other damages, is commonly believed to be responsible for the decline of forest dieback in the US. Eventually it was determined that the problem was more diverse. The diebacks in the US were largely due to a variety of regional factors, only some of which were related to sulfur dioxide pollution.[3] In Europe, research found that most of the dieback was a lagged response to root damage caused by severe droughts of earlier years,[4] although some forest areas immediately downwind of industrial facilities clearly seemed to be impacted primarily by emissions from those facilities. Gradually the dieback problems receded, and dieback concerns faded by the early 1990s. At that time, the issue of global warming took the place of acid rain as the most concerning global environmental issue, and many researchers were redirected to focus their efforts from forest dieback to climate change and its effects.

The Ozone Hole is another recent example of a global externality. It is due to the effects of the release of chlorofluorocarbons (CFCs) on the atmosphere. When released, this gas creates an "ozone hole" by depleting some of the earth's ozone shield. The shield, which is found in the troposphere, protects life on the earth from damaging ultraviolet rays.[5] The CFCs were produced for refrigeration by a few firms in only a few countries, but their release from old freezers did global damage to the Ozone Hole. The solution to this global environmental externality problem was a treaty, the Montreal Protocol (1987),[6] which called for a substitute refrigerant gas, hydrofluorocarbons (HFCs), phased out the use of CFCs, and provided compensation to those companies that discontinued CFC production.

Although largely successful, the problem has not yet been totally resolved because the HFC replacement for CFCs, while not highly destructive to the ozone layer, acts as a powerful GHG and has substantial effects on increasing the atmosphere's ability to capture the sun's energy, thereby increasing global warming. The approach for addressing HFCs is essentially an extension of the earlier Montreal agreement, but now focused on the removal of HFCs. It is moving to obtain widespread support for the gradual phasing out of HFCs over a thirty-year period.[7] An additional problem occurred in the form of CFC spikes in the atmosphere as late as 2018. This was apparently due to Chinese firms producing foam insulation for refrigerators. These firms claim that they were never apprised of the prohibitions against the use of CFCs.[8]

GLOBAL CLIMATE CHANGE

The current environmental problem, of course, is global warming, which involves an invasion of damaging agents—carbon dioxide and other GHG emissions—that are influencing the atmosphere. These are the agents responsible for the warming and its associated negative environmental damages. And these are agents of our own making, although for centuries and more they were viewed as nonthreatening. The damages are generated by almost all countries, all of which are producing GHGs. Thus, we have a variety of countries that, in the process of producing energy largely for peaceful and wholesome purposes, are inadvertently generating warming and associated damages. The damages impact both the mother countries as well as their global neighbors.

The issue here is one of global externality. GHGs are, in essence, a pollutant. Like other air pollutants, GHGs readily cross borders and indiscriminately warm both local and foreign areas, collectively generating warmer temperatures across the globe. But damages, particularly those of carbon dioxide, are not those of smog and dirty air. Carbon dioxide is colorless, odorless, and not destructive to human or animal health. Rather, its damage comes in the form of contributing to the modification of the atmosphere to promote global warming, climate change, and associated damages.

Uncoordinated mitigative action by individual countries, while helpful, is unlikely to be very effective in addressing the long-term global problem. Since it is cumulative mitigation that effects warming, unless done in concert, ordinary domestic pollution taxes or regulations by individual countries are not likely to be effective on the major sources of the damages. What is required is an over-arching coordination or centralized decision-making. For most ordinary goods and services, the market provides this function. However, lacking normal supply and demand forces for collective goods and therefore absence relevant prices, a market solution will not occur in the case of climate change. Instead, the world community has created a coordinating organization in the UN, which has chosen to undertake a collective action via treaties like the Paris Agreement. This collective action is designed to coordinate mitigation efforts to control climate change. However, at least one critical feature is missing in the UN—it is, largely by design, not a strong organization.

A GLOBAL EXTERNALITY

Global warming constitutes the world's largest global environmental externality. Although warming generates both positive and negative externalities, the consensus is that the negatives overwhelm the positives. Essentially every country is damaged in some manner. The object of policy is to find ways to discourage the GHG pollution that causes warming. A common approach is to use some government instrument— a "carrot" or "stick"—to discourage pollution. This could be by regulation, taxes, or subsidies. The objective is to place the costs of pollution on the polluters and thereby shift the responsibility and burden of pollution control on them. In the case of UN climate change agreements, the polluting countries accept responsibility by virtue of accepting the obligation to meet their carbon emission reduction targets.

The same rationale applies to GHGs as to a pollutant. If the damages are both local and international, local fixes alone are inadequate. Therefore, some degree of cross-country coordination is required to undertake global pollution control effectively.

ENVIRONMENT AND A ROLE FOR GOVERNMENTS

In addition to addressing externalities, governments can also have the role of providing the community with what are sometimes called public goods. The classic case of a public good is often identified as national defense, but could include highways, airports, and parks. Just as national defense is supposed to protect from the destruction and damages of as foreign power, the public good is the provision of the mitigation services, which provide for the mitigation of GHGs and the prevention of climate warming. The public good provided is a global good to offset a global environmental externality.

A second government responsibility is to protect its citizens from damages, including damages inflicted by environmental externalities that have not been prevented. These can be in the form of adaptive services, which might be viewed as governmental environmental services. Public goods in this context would not normally be produced by the private sector but would be public sector goods designed to anticipate and offset environmental externalities.

For an adaptation approach, the focus would tend to shift away from centralized decision-making and back to the individual state or even further decentralized to the local area and political unit. Thus, as in Chapter 4, sea-level-rise responses are unique to the circumstances of the individual seaside location. Dikes, dredging, new building standards, and relocation are all adaptations, but their specific applications would depend heavily upon the details and circumstances of each location.

POLITICS AND CLIMATE

An obvious example of a destructive change generated by warming is sea-level rise, but a host of other destructive changes can be expected. These damages are not confined to any one area of the sea but essentially effect the entire global ocean system. Since every country contributes only a small portion of GHGs to the global atmosphere and thus each has only a minor impact on the global ocean system, efforts by any one country to control the rise of the entire sea level through reducing warming has but a negligible effect. In response to this environmental condition, the direction chosen by the global community has been one of centralized decision-making via a UN vehicle. As with the Ozone Hole, the orientation has been to attack the externality directly by addressing the issue of fossil-fuel energy, carbon, and GHGs.[9] The focus has been on the reduction of fossil fuels and the substitution of renewables, and there has been a dramatic shift from high carbon emitting coal to lower emitting natural gas. These changes were initially promoted through the UNFCCC treaty and the Kyoto Protocol. But any global agreements must, by necessity, deal with political development, both globally and within individual nations.

Human societies have gone through several phases of political development. Early precivilized societies gradually evolved into early city states such as Ur, Kish, Babylonia, and Thebes. Over the millennia the city-state reemerged and gradually expanded, merged, and developed into national states. The more powerful of these states expanded into empires, whereby they controlled other states. Eventually, through history, these earlier political entities expanded only to eventually decline and fall.

Following World War II, Western Europe has moved toward a more integrated state, initially with a customs union arrangement, and gradually toward today's European Union, with relatively free trade and the increased mobility of labor. Internationalism grew through the vehicle of diplomatic interactions and military expansions as well as via international trade. Today, these activities generally continue to promote expansion and consolidation.[10] International trade today is driven largely by market forces, often within the context of overarching trade agreements under the World Trade Organization. Additionally, voluntary organizations such as the earlier League of Nations have tried to provide guidance via a diplomatic driver on political activities. Although the League of Nations failed to prevent a devastating World War, the United Nations has had more success, though clearly has been ineffective in preventing smaller but serious localized conflicts.

Although often criticized for its ineptitude and overly bureaucratic structure, the UN has found a place in the modern world. The approach chosen to address global climate change—mitigation under the umbrella of centralized direction—is consistent with the overall shift toward internationalism. This approach is consistent with the prevailing wisdom today, at least among the educated liberal class, and appears oriented toward increased collectivism. In the political realm, the lure of socialism continues to be strong. Despite the failure of the Soviet Union and economic collectivism in China, many of the young in the West find socialism surprisingly attractive. Even given the experience in smaller countries such as Cuba and Venezuela, where collective economic systems have wholly failed to bring economic prosperity and have been associated with repressive autocratic governments, the attractiveness of collective economics continues for many.

An adaptation approach to climate change would be less consistent with political centralization and UN-style control, and more in tune with a greater decentralization of global communities. Perhaps there is room for both in the struggle for surviving global warming.

CROSS-COUNTRY POLITICS

The politics of climate change have not been easy. Cross-country conflicts became quickly apparent. Early on, some far northern counties looked to climate change as a vehicle to bring them an improved competitive advantage for economic activities, particularly agriculture and forestry. More broadly, it was recognized that political responses to climate change, like the requirement of large carbon/GHG reductions, could generate huge shifts in certain fossil-fuel-intensive countries' comparative advantages, and thereby benefit producers situated to address more inexpensively a reduction in fossil-fuel energy, like the use of hydropower. An early manifestation was the competition within the negotiations over targets for GHG reductions in the Kyoto Protocol. This negotiation was particularly strong among industrial states, particularly the EU and the US, and was likely one of the reasons the US refrained from participation in the targeting.

The negotiation process appears to have the potential to become a disguised tool for income redistribution from industrial to developing. While redistribution may be desirable, it should be done with eyes open. We see today that the Paris Agreement, from which the US recently withdrew, has provided for an initial transfer of some $100 billion from wealthy developed nations to developing nations. Additionally, this proposed transfer is recognized as only the first of what is envisioned to be a number of substantial wealth transfers. It appears likely that obtaining the funds from wealthy countries could be difficult.

UN FRAMEWORK CONVENTION ON CLIMATE CHANGE (UNFCCC)

The global warming issue is similar to the CFC or HFC problem of the Ozone Hole discussed above, but much more complex. Not only are all countries negatively affected by the warming caused by

GHGs, but the source of the warming agents is essentially every country in the world, since all use some fossil fuels. For CFCs, the few producing countries have been compensated for ceasing production, and a substitute product, although with its own separate issues, has been found. Such a simple solution is unlikely for GHGs and global warming.

Nevertheless the basic approach for addressing GHGs has been similar in some respects to that of CFCs. For example, a centralized authority promotes the development of an agreement to address the polluting agent. In the case of climate, the overarching charter is the UN Framework Convention on Climate Change (UNFCCC), which has been signed by 195+ countries and commits them to act collectively to try to control GHG emissions. The US is a signatory to this agreement.

A problem with the FCCC is that, like many UN resolutions, it is nonbinding, so commitments are very loose. With so many participants, attempts at gaming the system[11] are common. The process of reducing the use of fossil fuels, for example, may reduce the international competitiveness of certain countries in some markets. One type of ploy is for countries to try to get commitments from others while avoiding seriously fulfilling commitments on their own. The first serious attempt to broadly apply specific GHG-reduction targets to individual countries was via the Kyoto Protocol. It met with partial success.[12]

Some countries were very heroic in their commitments and demands on others. The Europeans were particularly aggressive and then tried to benefit from inherent advantages.[13] Despite a prohibition against using credits for projects that would have been undertaken anyway, some countries tried to gain credits for GHG reductions from these types of projects. Germany, for example, was planning to renovate Eastern Germany's energy sector, which had been particularly intensive in carbon dioxide gas emissions under the Soviet regime before the fall of the Wall. The preplanned energy restrictions would greatly reduce their emissions and were to be undertaken regardless of the climate agreement. Thus, they gained credits, even though the assigned targets would have been met if they were not part of the agreement.

Great Britain was converting its energy system from coal to natural gas, a change that would dramatically reduce their emissions. They lobbied for credits for these projects, even though they had been envisioned for a substantial time. Other similar situations existed, especially in Europe. These conditions allowed some countries to meet apparently challenging targets easily and also allowed them to try to push others to aim for high emissions targets that were costly and very difficult to achieve.[14] A longer term effect could be for a country to meet a stringent target but only at a substantial energy cost disadvantage. This could upset the balance of country energy costs, putting countries at a competitive cost disadvantage for some of their products in world markets.

The US refused to commit to specific targets under the Kyoto Protocol, in part because of the absence of targets for major developing nations, particularly China. These countries had no targets but rather a loose commitment to move toward reducing emissions over a long period. However, even without a commitment to a target, the US did reduce its GHG emissions very substantially during the Kyoto period.[15] This success was due to the widespread substitution of low-GHG-emitting natural gas for coal. This substitution was driven primarily by new technologies, particularly fracking,[16] which lower considerably the costs of natural gas in domestic markets. The market simply substituted the low cost natural gas for the previously used coal. This procedure has contributed to a rather dramatic overall reduction in US GHG emissions.

In subsequent years, the use of Kyoto-like targets has not been repeated. Rather, the approach coming out of the Paris Agreement was a loose commitment to voluntary targets, as well as the transfer of monies from wealthy countries to developing countries to help finance the adoption of low emission energy.

THE POLITICS OF MITIGATION VS. ADAPTATION

The power given to the UN in its GHG monitoring function is large and adds to its strength in other areas like security. As we have noted in this volume, ways of addressing climate change can fit into one of two categories: mitigation and adaptation. These two approaches are associated with different degrees of political cooperation and coordination. As a global environmental externality that crosses international boarders, limiting global GHGs sufficiently to be effective in the mitigation of climate change requires a high degree of political cooperation and the participation of most of the world's major countries. The mitigation approach involves the recognition that no single country is a dominant emitter. Thus, political entities like the UN function as a coordinating institution. Such a situation, however, also can give rise to strategic free rider issues, where some countries could receive the benefits of GHG control even if they did not seriously participate in the GHG emissions reduction activities.

An efficient approach to use to try to control global warming would typically be determined by an assessment of the comparative costs and benefits of mitigation vs. an adaptive management approach. These types of decisions, although typically practiced in an informal manner, are common throughout life. Obviously, if mitigation were easily achieved at a low cost and low risk, it would be the sensible choice.

However, if the costs of mitigation are high and the chances of success modest, an adaptive management approach might also be attractive. In a case where mitigation is essentially impossible, as if natural forces were generating the majority of climate warming, the only alternative to address the natural portion would be some form of adaptation.

Adaptation has the advantage that it need not require a high degree of political cooperation nor centralized decision-making, since it involves countries addressing localized problems and local climate-related damages. Adaptation usually recognizes or acknowledges the warming event that may be occurring or is inevitable, then it tries to reduce or otherwise deal with the consequences. It is much like addressing a natural disaster in the absence of climate change. For example, suppose there is a hurricane, where we may know the event is coming but be unable to stop it. We may do things to prepare for it, and then, during and after the event, we can try to minimize the damages. This anticipatory approach is common today, as property and life are protected when we expect hurricanes. Note in this situation there is no need for close political ties to others, even if they, too, are impacted, as the responses are likely to be mostly domestic.

Most efforts that are directed to the prevention of the warming event are characterized as mitigation, but there are exceptions. For example, in the case of atmospheric manipulation designed to have reflective global affects, a single country with sufficient technical expertise and capacity could undertake activities to neutralize the warming capacity of the atmosphere.[17] Thus, a dominant country could, in principle, undertake a program of atmospheric reflectivity on its own without involving other countries.[18]

Earlier in this volume I discussed some global environmental issues that were met relatively successfully with both a centralized and decentralized mitigative approach. Acid rain and forest dieback, for example, turned out to be local problems amenable to local changes. In the US, although the dieback did not involve international issues, they could involve pollution crossing jurisdictional lines. The federal Clean Air Act eliminated that issue, since the same standard was imposed across jurisdictions. In practice, addressing acid rain now largely involves local decisions of addressing pollution in individual air sheds or within regional political units. This approach allowed sufficient air quality improvement to rectify the dieback problem as it was driven by air pollution.

The Ozone Hole, although not yet totally resolved, is a true global environmental problem that is being addressed with an international collective approach, called the Montreal Protocol. The Montreal Protocol involves a coordinated effort, under the UN, to prevent

CFC and HFC emissions. The UN Paris Agreement is structurally similar to, perhaps even a prototype of, the Montreal Protocol.

The climate problem is proving to be fundamentally more difficult to solve than acid rain, as it involves a situation where essentially all of the earth's countries are both users of fossil fuels and emitters of GHGs. Thus, they are experiencing damages related to warming even as they are contributors to the damages.

It is interesting to note that the general approach of UN-led mitigation is focused on human-created climate change, via fossil fuels and GHGs, but does not recognize natural warming. Response to natural warming is actually like the activities humans undertake today in response to natural events like floods, blizzards, earthquakes, volcano eruptions, and wildfires. The mitigation of GHGs is unlikely to reduce natural-sourced warming. Adaptation can be associated with an anticipatory plan, that might, for example, involve a collection of requisite equipment and other resources, and the addressing, as best possible, of natural climate change that also threatens humans. Climate change adaptation management thus would add an additional dimension of natural disturbance to the human environment.

THE UN AND CLIMATE

After World War II, the UN was designed primarily to address international issues that extended beyond the borders of any one country. Its predecessor, the League of Nations, was created out of the ashes of World War I but had failed to forestall World War II and ultimately collapsed. Toward the end of World War II, the international community, led by the US, moved to try again. The UN was created along with a number of associated UN agencies. These included organizations such as the IMF and the World Bank, designed to deal with financial and development issues. Other UN agencies were crafted over time and gradually the UN moved into environmental areas. The UN Environmental Program was created in 1972.

Simultaneously with the creation of the UN, organizations in Europe were established to foster cooperation and coordination. Initially, an important task was to facilitate Europe's reconstruction, trade, economic activities, both within Europe and among other developed countries worldwide. Subsequently, the organizations became more powerful and more integrated and now include the European Union (EU) with a common market as well as a common monetary system featuring the euro as its currency.

THE UN INTERGOVERNMENTAL PANEL ON CLIMATE CHANGE (IPCC)

The UN Intergovernmental Panel on Climate Change, UNIPCC, was created in 1988 and given the charge of undertaking a continuing review and monitoring of the effects of GHGs on global temperature and climate. Its major task has been the preparation of periodic reports or Assessments. An Assessment Report, consisting of the outputs of three working groups— physical science, impacts, and mitigation—is prepared every several years. The Assessments are essentially periodic reviews of the state of climate-related research in the effects of human GHG emissions and their impact on the global climate system. An Assessment is undertaken about every seven years or so, with the latest in 2014.

Rather than developing new research, the charge of the IPCC is to undertake a huge multifaceted literature review of climate science, including physical, social, and economic literature. Projections of future climate are a part of the IPCC charge. The general approach is to divide that task into the three working groups and produce the Assessment around them. The topic areas are further subdivided into a number of sections, with each becoming a chapter. A group of researchers (lead authors) with expertise in the topic areas within the broader scope of the chapters is selected as a writing committee. Two co-chairs, called convening lead authors, usually one from the developed world and the other from a developing country, are selected to oversee and coordinate the effort. This basic writing committee is assisted by a number of contributing authors, who are asked to prepare sections on their respective areas of expertise within the broad topic. Ultimately, these contributions may be integrated into the text at the discretion of the committee authors. The contributing authors create their contributions independently. For each Assessment, the entire set of chapter lead authors meet a few times over a period of roughly two years to try to coordinate both within and across chapters. Much of the writing communication is done via electronic media.

Ultimately, the advanced draft chapters are presented to the participating countries, usually at a multicountry meeting, where discussions occur and disputes over the materials often arise, with the political views of the countries often pressed upon the researchers. This sometimes necessitates pushback by the chapter chairs to try to minimize the extent to which the countries' political agendas find their way into the Assessment chapters, which are intended to describe research findings rather than promote political agendas.

In addition to the Assessment, other documents are produced, e.g., Summaries for Policy Makers (SPM) and Technical Summaries of the Working Groups. These focus on background issues and/or summarize the parts of the much larger Assessment and allow organizers substantial latitude for emphasize some aspects of the Assessment while downplaying or ignoring others. In my experience, the senior IPCC Secretariat representatives use a strong hand in selecting the topics to be discussed and how they were portrayed in the SPM. Many authors participate in more than a single Assessment chapter, being involved in different parts of the Assessment over time as an individual's role would often be reshuffled.

MY PARTICIPATION IN THE IPCC

Over the years I was invited to participate in some capacity in three of the early Assessments: Numbers Two, Three, and Four, published in 1996, 2001, and 2007 respectively. Subsequently, another Assessment has been published in 2014, and a further is due in 2019.

The first Assessment in which I participated was the Second Assessment, prepared and completed in the mid-1990s. My modest contribution in the Second Assessment consisted of preparing two small sections that were subsequently incorporated into two different chapters. I played a much larger role in the Third Assessment Report, completed about 2001, where I served as a co-convening lead author (co-chair with Pekka Kauppi of Finland) of a chapter. As a co-chair, I was involved in finalizing our chapter and defending it in a forum of political participants that met in Ghana. I worked with the Secretariat on finalizing related documents, including the very influential Summary for Policy Makers.

In the Fourth Assessment, I again played a more modest role, participating as a de facto lead author in the preparation of the chapter assessing the impact of climate change on agriculture and forests. The behavior of the IPCC regarding my involvement in the fourth report revealed the political nature of the choice of participants. The usual approach was for candidates to be nominated by governments and then chosen by the IPCC Secretariat. In this case, I chose not to be a candidate as part of the original selection. However, one member with my specialty in the chapter group had dropped out early in the process, and the chairs of that chapter contacted me to see if I would be available as a substitute. Upon my concurrence, they recommended that I fill the position. However, the Secretariat refused to appoint me as a lead author on the grounds that I was an American and they had determined that there were already too many Americans. This was not a subtle move, as the Secretariat member unabashedly articulated the rationale. I recall counting the number of American participants and believing that although large, it was not overwhelming, particularly when compared with European participation as a group. In any event the determination was that I would be appointed at a lower level of involvement as a contributing author.

This solution presented problems in that there were no travel funds for contributing authors from the US to attend the four international meetings involving lead authors. I was provided with "special" funds from a US agency to attend one of the international meetings, and I was otherwise involved through the writing process via Internet and email. However, the one international meeting I attended turned out to be most productive in that an essentially complete final draft was prepared. For these collective efforts on three Assessments, I was later awarded a share of the Nobel Peace Prize (2007) that was received by IPCC.

A FEW CONCERNS ABOUT THE IPCC

As I became more involved in the IPCC, a number of features about the process disturbed me. First, the charge given to the IPCC, as I understood it, was not to examine the phenomenon of climate change per se so much as to assess the effects of human activities and especially carbon and GHGs on the earth's current warming. This is a far narrower and decidedly less comprehensive task than that of investigating the broader warming phenomenon. A broader assessment would involve a focus on not simply of one hypothetical cause, human impact, but more on an understanding of the entire climate system. A narrower focus on human-generated impacts, I believe, has led to a lack of a fuller investigation and understanding of the relation between the impacts of GHGs and the natural warming that has occurred.

The fact that other efforts to understand the broad phenomenon, including both human and natural forces, while addressing correcting activities were given relatively little attention has resulted in only a very modest effort being devoted, for example, to geoengineering possibilities. These include a number of engineering schemes (see Chapter 5), e.g., atmospheric reflectivity or Earth albedo, where the atmosphere might be modified to enhance the reflective effects, reducing the amount of the sun's heat captured.[19] Although research was being undertaken on climate change over earlier historical and prehistoric periods, these earlier experiences did not seem to play a serious role in the interpretations of the Assessment. I note here that in the Third Assessment, our chapter had the responsibility of discussing geoengineering. However, we had little expertise in this topic among our lead authors and the section was highly perfunctory.

So the focus of the Assessment was largely on the effects of human GHGs. Although this was appropriate at one level, an understanding of human activities in a broad context that includes natural variation would seem to be more appropriate. There was an acknowledgement that other forces also promote warming, but these tended to be persistently downplayed and a fuller analysis is absent. The analysis focused solely on approaches to offset warming from GHGs and did not try to address the full fundamental causes of warming. Behaviorally, the approach simply assumed warming was due to human GHGs with little attention given to earlier natural warmings.

In the Fourth Report, we were given the task of discussing agriculture and forestry. However, specific temperature scenarios, which are useful, were not provided, despite our request. Rather we were responding to a generalized warming. I note here that the general view is that warming will be greater as one approaches the polar areas. Obviously, this has implications for locational aspects of agriculture and forestry, but we were never able to consider those in our chapter.

An additional problem with the IPCC was the makeup of the writing committee. Early in the process some participants were inserted into the chapter committees with only modest technical expertise in the subject of the chapter, but typically with strong ideological opinions. I was concerned that they were placed there to try to insure "politically correct" decisions in the papers.

The participants had been recommended by different countries, presumably for their technical expertise, and then selected for chapters based on their specialty and country affiliation. There was a definite effort to have developing countries well represented, as well as a geographic balance among participants. In the preparation of the Third Assessment, where I was particularly involved as a convening lead author, some chapter groups developed into ideological divisions as some countries appeared to promote individuals who were more environmental advocates than scientists or technical professionals. However, some genuine technical differences did emerge.

The problem of advocates had become so broad that a large number of legitimate participates, mostly Americans, called for a meeting with members of the Secretariat to complain. Advocates were not desired since a scientific approach should be free from predetermined objectives and instead driven entirely by the science. This is the charge of the IPCC. The outcome of the meeting, as I recall, was for the environmentalists to withdraw from participation in the writing process. Gradually, these individuals eventually reinserted themselves into the committees. In my group, we had one particularly vociferous advocate participant. This advocate infiltrated our chapter meetings and eventually was asked not to speak and was subsequently ignored. In the end, however, the problems were not insurmountable, and overall some useful work was accomplished.

A major concern I had throughout the process was the strong advocacy bias of the Secretariat's management staff. Commitment is clearly desired in a staff. However, being a scientific effort, I thought objectivity was also important. Much of the staff was strongly committed to generating a document that would promote action rather than one that would promote thoughtful assessment. I note one meeting in preparation of the Summary for Policy Makers. When it was suggested that a paper be cited that suggested delaying actions in order to more fully educate the readers about mitigation possibilities, a senior staff person stated, "No, that will just give them (the countries) an excuse to delay." This citation was omitted.

Part of the issue involved the question of how quickly the world should proceed with mitigation. One view was to spend more time understanding the warming phenomena and then proceed a careful manner. This would provide time for consideration of approaches other than mitigation, perhaps adaptation and geoengineering. The other approach was to charge ahead in trying to address the warming problem as we understood it. The second view was that of the Secretariat and clearly carried the day.

A BIAS TOWARD MITIGATION?

During my early research on climate in the late 1980s, I saw a strong bias toward a mitigation perspective in the treatment of an RFF project in which I was involved in the late 1980s and early 1990s.[20] The project, headed by Norman Rosenberg at RFF, was funded by a major government agency. It was designated to investigate the likely impact of a return of the dust bowl climate of the 1930s to the agricultural states of Missouri, Iowa, Nebraska, and Kansas (MINK). In essence, such a climate would be of a type that might be anticipated, should global warming continue. The project was focused specifically on how these four states might adapt. However, although clearly a climate-change study, we were instructed not to present it as such.

Our funder was uncomfortable with that designation and preferred to treat it as a simple weather project, not as climate-change related. This project generated quite interesting results as to the substantial potential of adaptation to climate, even in a mostly agricultural environment. The results were presented at conferences, in published papers, and in a book. Nevertheless, upon successful completion of the project, we prepared a logical follow-on that was not funded, even as the number of mitigation projects addressing climate-change issues was greatly expanding.

This response could have simply been a coincidence. However, during the decade of the 1990s, there was an increased resistance to adaptation projects. I recall being counseled not to use the term "adaptation" in the context of climate discussions, and indeed we did not characterize the RFF MINK project as a global-change project, but rather as an agricultural project.

Why might politics lean toward mitigation? There are a number of reasons. First, there is a tendency to collective action, especially within dominant capitalist economics. This can be seen in the increasing coordination of the western economies. This tendency is reinforced by the existence of the UN, which by design is suited to address collective issues, such as international environmental issues, with a top-down approach inherent in global mitigation. Second, given the relative success of earlier attempts to address environmental issues associated with the Ozone Hole, the selection of a very similar approach, again through the UN, is not surprising. Third, there is the powerful tendency to view warming as generated simply by GHGs. In that case, a simple mitigative approach seems appropriate.

Clearly, if the climate system were as unsophisticated as the simple carbon/GHG model, then mitigation would be the most obvious approach. However, some reasons to consider adaptation include the following: The mitigation approach is very costly and high risk; high risk because there is always a significant chance that the climate process is more complex and that reducing the growth of GHGs may be unachievable or inadequate to stop warming. Another risk is that efforts would be put into stopping a problem that, it could turn out, was much less severe than commonly believed. Even if the focus is limited to human-induced change generated by GHGs, and any natural forces are ignored, the dominant mitigation approach involves a comprehensive remaking of the global energy system away from fossil fuels and to renewables. If all of the climate change were due to human GHGs, then while the costs might be high, the long-term risks would be modest. As stressed in this volume, the earth's climate history shows a high degree of natural variability, and even the IPCC acknowledges that other factors, undoubtedly natural, contribute to climate change.

If the potential for natural forces to promote climate change is factored into the assessment risk of mitigation, the degree of risk with a mitigative approach increases substantially, since GHGs apparently did not play a significant role in historical warmings like the Medieval Warming. Natural warming does not appear to be dependent on GHGs,[21] and thus could proceed even if the mitigation of GHGs is highly successful in reducing atmospheric gases. So, humans could make the transition to a largely renewable GHG-free energy system without substantially reducing global warming problems.

Adaptation involves a continuation of what humans are doing to anticipate and adapt responses to major catastrophes—weather events such as hurricanes, declines in agricultural productivity and management, and preservation of forests and natural terrestrial ecosystems. Many adaptation approaches are far less costly than reprogramming the world's energy systems. As outlined in Chapters 4 and 5, there are a host of viable anticipatory and adaptive approaches. Some have argued that other approaches have not yet been well-developed. However, sometime back, Wigley, Richels, and Edmonds[22] argued that the optimum approach to dealing with warming may be to spend some time, perhaps a couple of decades, improving our understanding of the problem and developing more efficient technologies before trying to address climate change head-on. If projections are accurate, today we have much less time to address the problem. Although there are many unknowns associated with geo-engineering and other solutions, there have been several decades to work out these questions. The costs of geoengineering solutions are very much less than those associated with GHG mitigation.

The risk is that mitigation efforts could be of limited success, since they are applied in an environment where other forces, particularly natural climate variability, are outside the reach of mitigation and any other human control. Thus, by this consideration, it would appear that a reasonable assessment of risk would lead us to address climate change by adaption approaches.

Nevertheless, the global political process has leaned heavily toward using a mitigation approach when attempting to address climate change. Only recently has it become clear that mitigation probably will not, or cannot, restrain temperature to within what have been determined to be acceptable levels, resulting in an increasing emphasis on an adaptation approach.

CONCLUSIONS

The global political process has worked with the global community to address the climate issue using a top-down, centralized approach. This approach, administered by the UN, follows its earlier success in addressing the "Ozone Hole" problem, with a similar institutional arrangement. It is aimed at reducing or eliminating GHG warming by dramatically reducing the use of fossil-fuel energy and substituting renewable energy. Such an approach, what this book has called Plan A, involves a great deal of political and economic coordination.

The UN is the coordinating organization under the FCCC, with the IPCC functioning as a monitoring agent for climate research. Mitigation was chosen due to the commonly accepted view that the climate problem is driven largely or even solely by GHGs. The mitigation process is not yet nearly complete. Fossil-fuel energy continues to dominate, and the modeling of future trends raises serious questions as to whether the carbon targets can be met in a timely manner. More and more assessments indicate serious problems with meeting the GHG and temperature targets using only mitigation.[23] Questions, rarely discussed publicly in the political arena, about the role of natural climate warming point to potential complications in the viability of the mitigation approach. Alternative approaches are available that might be better suited. These alternatives generally can be applied locally, and so require less centralization. Individual countries are increasingly undertaking adaptive approaches.

CHAPTER 7

PLAN B AS INSURANCE

once moved from Washington, DC, to a western state where I bought a home on the ridge of a mountain. As I assessed my new residence, I became concerned about the possibility of earthquakes and earth movements. After all, there was my house on the edge of a cliff. I contacted my insurance company in DC and asked about earthquake insurance. My agent said, "Oh, you don't need that insurance. I am originally from that area and they don't get earthquakes or earth movement (unlike your neighboring state, California)." I replied, "since the risk is so minor, the insurance should be very cheap." Indeed, it is cheap. I now can sleep more easily with a low-cost insurance policy protecting my family and my house against a very unlikely, but potentially catastrophic occurrence.

The world is full of risks and uncertainties. Insurance is a vehicle for protecting oneself against these eventualities by moving into a "pooling" situation where those risks and uncertainties are shared. We all pay premiums so that those who are injured can receive some compensation. One can think of the options as follows: A risk exists, say, of fire destroying a building. The owner can choose to insure the building. If there is a fire, the decision to purchase insurance will pay off. If a fire never occurs, a payoff will also never occur, and the premium will have been unnecessary. (The only advantage of the insurance may be that the owner can sleep better at night.) Without insurance, the full risk of the loss falls upon the owner. With insurance, some portion of the risk, perhaps all, is shared by the insurer.

Global societies face the same type of uncertainties. Suppose the globe incurs the costs of mitigating GHG emissions and converting global energy systems away from fossil fuels and toward renewables, and the risks and extent of global warming were radically exaggerated? Suppose radical energy conservation was not necessary. Then we might see most of these changes as having been unnecessary, at least in the time frames in which they were undertaken. We have bought and paid for an insurance policy that was not necessary. However, if warming is as serious a problem as portrayed in this volume, the insurance of eliminating fossil fuel use and conversion to renewable energy was probably justified, assuming it was successful in mitigating warming. However, if the mitigative approach, while rightly directed, was inadequate to control warming, the rationale for adding a Plan B would be overwhelming.

There are a number of ways to apply insurance to the risks of climate change. The most direct is to apply a traditional type of insurance to directly provide for compensation against the type of damages that are commonly associated with climate change, for example, the damage from flooding. At the other end of the insurance perspective would be to provide the insurance in physical protection. For example, a property owner could build a seawall to reduce risks of damage that would occur in the case of a sea surge, which would now be more likely due to sea-level rises. Another example would be found in the government providing the insurance in the form of a government-financed protective seawall.

Perhaps a more common approach would involve government-initiated policies, like Plan A policies that undertake massive, coordinated, multigovernmental GHG mitigation. The Paris Agreement can be viewed as an attempt by the global community to provide insurance against the prospect of massive damages coming out of a warming episode.

There are a host of intermediate approaches, such as some form of insuring activities found in adaptation; for example, the insuring of dikes. A California state senator has suggested an interesting insurance application. The plan is for business, local governments, and perhaps even the state to pay premiums to maintain and protect natural buffers as insurance against sea-level encroachment.[1] This will promote natural barriers such as wetlands, forests, and other buffers to reduce climate-associated damages.

HOW COSTLY ARE CLIMATE DAMAGES?

Various estimates have been made of the costs, in financial terms, of climate change.[2] These costs are found in damage assessments associated with sea-level changes, energy costs, and ecological disturbances. An estimate commonly used by the US government is one where each ton of additional atmospheric carbon dioxide costs society (in damages) about $36 per ton.[3] Humans release roughly four thousand million metric tons of GHG annually,[4] or forty billion metric tons.[5] If valued at $36 per ton of carbon dioxide equivalence GHG, this converts to $1440 billion or $1.44 trillion in global damages annually.[6] However, some more recent studies have estimated the costs at a considerably higher rate of $220 per ton.[7] These higher damages are usually found when focusing on extreme weather and climate-related events such as hurricanes, floods, and extreme temperatures. These estimates are of the long-term present value of costs of additional damages over time associated with one additional year's worth of higher temperatures.[8] The estimates of the value of climate damage,[9] especially through time, depend critically on the choice of discount rate.[10] In general, the higher the choice of discount rates, the lower the total damage costs.[11]

William Nordhaus, of Yale University, a 2018 Nobel Prize winner in economics, estimated the optimal carbon tax at $30 per ton of CO_2. This is the tax level that equates with the monetary value of climate damages estimated to be caused by carbon emissions. However, he noted that the range of the optimal tax could be from $6 to $93 per ton.[12] Subsequently, economist Robert P. Murphy estimated that with the addition of abatement costs, which would be required to abate existing, new, and excess GHGs in the atmosphere, the total cost of addressing the entire GHG climate event would be $22.6 trillion.[13] Note these are the total global present value costs for the entire multiyear warming event, not simply for a single year's worth of damage estimates.

The occurrence of these costs demonstrates the large social risk that societies face with continuing climate change. When functioning in situations that involve substantial risks, a reasonable approach is to try to protect oneself. We can view the program of GHG mitigation as a type of social insurance provided by governments and financed largely by the taxpayer. Note that insurance protects against events that may or may not occur and also provides coverage for events certain to occur, such as a death payment.

INSURANCE PROTECTION AGAINST CLIMATE CHANGE

In human interactions insurance is an institution that allows for the transfer of risk from an individual to a pooled group of risks by means of a two-party contract. The insured party obtains specific coverage against an uncertain event for a smaller but certain payment.[14] Typically, an insurance provider assumes a number of risks it pools in return for payments. The individual's payment is determined by the nature of the risk, the coverage, and the compensation agreed to. But insurance coverage varies considerably. It may provide compensation upon death to provide for survivors, medical coverage in case of an illness, driver's accident insurance, fire insurance, flood insurance, and so forth. There is also the concept of self-insurance (self-financed) whereby the principal and agent are the same entity and an individual (or organization) chooses to bear the financial risk of a negative event.

Although we typically think of insurance in terms of financial risks, the concept is broader. Dams are built, in part, to reduce the risks of injury from flooding. Stoplights reduce the risks of accidents between automobiles and pedestrians. Life guards reduce the risks of drowning at public beaches. These are all examples of socially financed services that reduce specific risks to the population. What these investments do is reduce the probability of negative events, even if the probability is not reduced to zero.

In the case of climate change, Plan A can be viewed as a form of self-insurance undertaken collectively by the nations of Earth. One question that needs to be asked is: "How much insurance do I need to increase the protection to me (and my family) from the risk of this event?" In some cases where we know an event is inevitable, as with death, the purpose of the insurance is to compensate the survivors for the loss. In the case of climate change, there is uncertainty between the rate and level of climate change and the effectiveness of mitigation activities. Most analysts believe that current levels of mitigation, which may be interpreted as the level of insurance, are inadequate to properly cover the major risks and therefore inadequate to avoid major damages. To address global warming, societies are providing investments to promote the mitigation of climate change and thereby to protect their populations from the risk of damages associated with warming.

If we have confidence that the type of insurance embodied in Plan A is appropriate, it then becomes a case of providing the right amount of insurance to minimize and/or compensate for damages. In the case of global warming under the Paris Agreement, the judgment is that by preventing temperatures from rising above 1.5 degrees Celsius, the more severe damages can be largely avoided. However, the ability of Plan A to meet this target in a timely fashion is an open question. As noted elsewhere in this volume, many projections and forecasts have suggested that current plans, essentially current insurance levels, are nowhere near adequate to meet that temperature objective.

The current approach is also expensive. We know, for example, that the Paris Agreement is targeting $100 billion a year over many years to finance the renewable energy needs of developing countries. This cost does not include any of the costs for developed countries to make the transition from fossil fuels to renewables. These costs include those of the shift away from fossil fuels that will necessitate the early retirement of costly plant equipment and its hasty replacement.

Although it may be argued that at some time in the future of humankind such changes would eventually need to be made, it has also become apparent that there continues to be vast quantities of remaining, low-cost, exploitable fossil fuels. Technological innovation, such as hydraulic fracking,[15] will continue to provide exploitable resources, as will technologies that provide lower-cost access to other accessible fossil fuels. These energy resources have reduced energy costs by delaying the need for renewables into the more distant future.

The above discussion suggests that despite investing money in "climate insurance" in the form of mitigation, the global community may nevertheless fail to adequately control human GHGs and thus face the daunting challenge of dealing with "extreme warming." Further complicating this issue is the role that natural climate variation may be playing in the warming. Thus, controlling GHGs simply may not be adequate to control warming. Prudence would dictate having a backup insurance. Much of this volume makes the case that the present insurance policy of Plan A is inadequate. Our suggestion is for Plan B to serve as that backup or supplemental insurance.

ADDITIONAL INSURANCE CONSIDERATIONS

There are a number of considerations when purchasing traditional insurance. How much insurance is appropriate for the event to be covered? Is the cost of the insurance too expensive? How much confidence do I have that the company will pay? Are the costs and benefits actuarially sound? Is there another company or type of insurance that is more cost-effective? Can I take some types of preventive actions to lower my risks, e.g., have fire extinguishers at certain high-risk locations on my property to reduce risk of fire damage? If society is viewing mitigation activities as insurance, the same types of questions should be asked.

Note that in the case of global warming, the investments in insurance thus far have been highly skewed to mitigation and focused entirely on warming caused by human-generated carbon and GHGs. However, there are many types of considerations when signing up for insurance. The set of decisions go something like this. Do we think a warming is likely? If so, what is the cause? If the cause is human GHGs, can we slow or prevent it? What are the preventive options?

BACK TO MITIGATION VERSUS ADAPTATION

Throughout this volume, I have contrasted two basic options: mitigation and adaptation. Obviously, with other things being equal, mitigation would appear to be preferable. However, is it achievable in principle and in practice? Questions have arisen about the ability of society to offset the buildup of GHGs in a timely fashion, driven heavily by the demands for fossil fuels from the developing world, including China and India. Additionally, the question of the effects of the ignored forces of natural climate variability could further exacerbate the problem of climate control. These uncertainties about the ability of mitigation to adequately address the problem raise the question of what is the backup approach, what is the insurance if mitigation, fails?

Note, however, that Plan B is designed to be more than a backup. It is intended to be simultaneously applied *with* Plan A, not in place of it. Adaptation can be viewed as two distinct and rather different approaches. The first approach involves addressing the existing stock of atmospheric GHGs and the warming associated with it. Adaptation, or adaptive management, involves first an anticipation of the damaging event and an understanding of the driving forces associated with the damages. Second, it involves an undertaking of steps to minimize the harm done by the damaging forces. As discussed, for sea-level rise this might involve the relocation of some roads, the replacement of older bridges with new and higher ones, and the use of more well-placed dikes. To deal with more extreme anticipated surge issues, improved water storage and drainage facilities would be appropriate.

The second dimension of a Plan B involves actions to neutralize the earth's warming, no matter what the cause. This could be undertaken by addressing the atmosphere so as to neutralize any changes in its effect on warming, as in the use of geoengineering approaches.[16] Different approaches that capture atmospheric carbon could be relevant.

It appears that more effort can and probably should be put into research related to the development of a workable geoengineering fix. As noted in the geoengineering chapter,[17] there are a variety of technologies and techniques that appear to have promise. Assessment of the strengths and limitations of them might be used as a filter to determine where research efforts appear most likely to be fruitful.

Note here that the argument of this book has never been that Plan B or the geoengineering aspects of Plan B be implemented in isolation or as a complete substitute for GHG mitigation. Rather, the intent is that Plan B be used as insurance against any failures or shortfalls of Plan A. Even some experts have argued that "climate engineering does not appear to be an alternative option, although it could be used possibly to complement mitigation."[18]

Some observers have resisted the use of geoengineering to neutralize the warming effects of the atmosphere. They note, for example, that there may be unintended consequences to the technology, even if warming is controlled. Well, yes, but proponents of geoengineering can argue that humans have already affected the atmosphere and the climate due to the increase in carbon dioxide. Future damages have been forecasted due to climate change. In the absence of clearly identified, unintended damages due to climate-control programs, we need to move ahead with a supporting level of effort or backup insurance.

REINSURANCE

In the financial insurance world, insuring companies have a backup in the form of reinsurance companies. These are companies with large resources to financially back up ordinary insurance companies. If a firm wants insurance against risks of damages due to warming, these policies are likely available to augment the retail insurance company. However, with climate change we are referring to risk protection for the entire world vis-à-vis climate catastrophes. In this context, the world community can be viewed as both the insurer and reinsurer. If the world moves into an era of larger and more recurrent risks of damage and catastrophe associated with climate change, the costs of adjusting would increase. More costs would be incurred to try to reduce the damages and thereby lower insurance. In such a world, the cost of insurance to the insured would rise, as would payment costs from the insurer to the insured for damages. There then becomes the question of whether the insurance company and industry can cover its now greater liabilities due to the likelihood of greater damages.

There is also a question as to how to guarantee that sufficient funds will be available for very large losses that could result from catastrophic events that might be associated with climate change. The financial pools need to be large enough to assure clients that the funds will be sufficient and available to at least cover a part of the losses. Today, the funding source for reinsurers is being expanded to include catastrophe bonds. These bonds essentially package insurance risk as debt. The funds from the sale of the bonds are added to the premiums paid to the reinsurer to enhance the size of the financial pool. The bonds, which attract investors by paying fairly high yields, do not have to be repaid in full or at all if a disaster stipulated in the contract strikes. Thus, the risk associated with the disaster is now shared with the bond holder. More funds are also made available to the insurance pool from investors who are willing to bear the somewhat higher risks of the catastrophe bonds in the hope of receiving those higher yields.

CONCLUSIONS

Insurance can play a role in addressing the risks of climate change, whether in the traditional role of compensating for financial values lost or as a deigned protector of property and lives. Just as "Plan A, mitigation" provides a social safety net, so, too, does "Plan B, adaptation" provide a supporting role. Geoengineering can be an important part of that support by reducing the probability of uncontrolled damages.

CHAPTER 8

WHERE FROM HERE? NEXT STEPS AS WELL AS SUMMARY AND CONCLUSIONS

THE ARGUMENT

T he argument of this book is simple. It accepts the now conventional Gore-type global warming model in which human
release of GHGs from fossil-fuel energy are likely the most important
factor contributing to the current global warming. It also accepts the
empirical estimate that global temperatures are rising and changes
to the global climate system and other facets of the global biosphere
are underway. This volume has reviewed many of the forecasts made
regarding changes due to global warming and accepts the view that
major changes, perhaps draconian ones, can be expected in the
absence of suitable control of GHGs. It further accepts the view that
current mitigating activities are insufficient to prevent major large-
scale future climate-generated damage.

Some have suggested that the solution is easy and the difficulties largely political. For example, Nobel Award winning economist
Paul Krugman argued in the *New York Times* that the costs of renewable energy have fallen so dramatically that "there is no reason to
believe that decarbonization would impose any significant economic
cost."[1] However, researchers Daniel Raimi and Alan Krupnick of
Resources for the Future note that Krugman "leaves the readers with

the impression that decarbonization would be cheap and easy if it weren't for entrenched fossil fuel interests." They argue that "the sheer scale of the energy system means that even the most rapid transition (to renewables) would take many decades."[2] Furthermore, the experience of Germany and other countries that are heavily dependent on variable energy sources such as wind and solar have found that backup and energy reserve requirements increase substantially as variable energy sources increase their share of total energy supply. Energy must be provided even if the weather is windless and cloudy.

In order to address the impending global warming disaster, this volume suggests a backup strategy to supplement the current mitigation approach. That strategy is one of climate adaptation. As discussed in the text, the strategy involves an approach of anticipation and adaptive management that also involves offsetting or neutralizing carbon dioxide's warming effects. Different approaches are available, including techniques using aerosols or cloud brightening to reduce the warming effects of GHGs.

The discussion also addresses some unexplained features of warming. An important argument is that much of past historic and ancient climate change is not explained by GHGs. I note that none of the earlier warmings, historic and prehistoric, were associated with human activities, and many, if not most, of the earlier warmings occurred without any increases in GHGs. This is clearly true for earlier warmings in the current interglacial period. However, the evidence unmistakably indicates that the earth has experienced a multitude of climate variations over literally billions of years, including relatively recently. These include major glaciations lasting hundreds of thousands of years as well as more modest climate changes such as the Medieval and Roman warmings, and the Little Ice Age, which ended only in the nineteenth century.[3]

A host of forces drive climate change. The glacial ice age periods are usually attributed to orbital forces, but the earth has also experienced modest and very short warmings from sunspot cycles. There is additional historical and ice-core evidence for the Medieval type of warmings that occurred during the interglacial periods as recently as within one thousand years. These have been preceded and followed by cooler periods, such as the well-documented Little Ice Age.[4] In a global climate system, with climate driven by a multitude of forces, some not currently well understood by science, it is not inconceivable that natural warming could also be occurring today together with human-caused, GHG-driven warming. In addition, the failure of scientific studies to fully explain all or most of the current warming suggests a gap that leaves a role for a natural warming, the implications of which I have considered in this volume.

Although accepting the essence of the GHG theory of climate change as an explanation for much of the global warming that Earth is currently experiencing, that view is incomplete.[5] I have identified some areas of weakness in the contemporary climate change theory and have shown where the theory is not entirely consistent with current empirical evidence and is inadequate at explaining earlier warmings.

Our discussion has examined earlier warming periods and taken note of some inconsistencies between their warmings and current human-driven GHG warming theory. Earlier warmings and climate changes, such as the Medieval Warming (1000–1300 CE) and the subsequent Little Ice Age, both of which are well documented in the historical record and in ice-core evidence, have occurred in the absence of the involvement of humans or rising GHG levels. Although these earlier warmings are still not well understood and have not been integrated into the GHG warming theory, they are part of Earth's climate history and as such must be part of a system of natural climate variation. Given this fact, the emphasis of the IPCC and other research bodies on the human impact on global warming could explain the failure of science to build these earlier natural warmings into a comprehensive account to explain current increases in atmospheric temperatures. The challenge for humans today is to control the warming, preferably by its reducing or eliminating fossil-fuel GHGs, augmented by successful adaptive measures.

I have also examined current projections and forecasts of future warming and the effects of such global mitigation efforts like the Paris Agreement to control future warming. I find overwhelming evidence that the forecasts of warming will overpower and outrun the forecasts of mitigation control. The implication is that if both sides of the forecasts are anywhere near accurate, the potential for major damage appears likely to overwhelm effort to curtail or stop human production of GHGs. Additionally, I have examined adaptive tools to address warming. These tools include both adaptive management and geoengineering measures. I examined and discussed some of the applications of these tools and suggested that they need to be critically applied along with the existing mitigation program. I argued that together these two approaches will enhance our ability to deal successfully with the global warming problem.

WHERE ARE WE NOW?

Where are we now, at the end of the second decade of the twenty-first century and some thirty years into a period of wrestling with the realization of a major climate change in our era? The world has moved ahead on two fronts. First, it has begun to seriously address the warming issue with the help of a centralized international organization in the form of the IPCC of the UN, around which to organize its efforts. It has generally accepted the guidance of the UN, with substantial input from several of the major developed countries, including the countries of the EU and, hauntingly from time to time, the United States, to address warming in a coordinated and centralized fashion.

Organizationally, the UN has been charged with monitoring the warming phenomenon and related research and forging a response to the warming. It began its response by creating the IPCC to oversee much of this centralized effort, including the collection of information, the periodic publication of the Climate Change Assessment, and the creation of related international agreements, such as the Kyoto Protocol and the Paris Agreement. The first effort, the Kyoto Protocol, has been reasonably successful in getting countries, particularly the developed ones, to focus on this issue by allocating to them emission reduction targets, which was to some extent followed up by emission-reducing activities in many of these countries together with voluntary monitoring to confirm compliance. In the subsequent Paris Agreement, developing countries also committed to voluntary targets.

Second, the world has accepted the working assumption of sectors of the scientific community that generally support the claim that the global temperature is rising, which is likely due to human actions. This human effect comes through the intensive use of fossil-fuel energy and the emissions of carbon dioxide and other GHGs. The world has also developed a set of technologies that offer the promise of replacing GHG-emitting fossil fuels with wind, solar, and other renewable alternatives.

Today, many developed countries have shown an ability to stabilize and even reduce their volumes of GHG emissions. Associated with this accomplishment is the development of important innovations in both the renewable and fossil fuel sectors. A major source of the improved (lower) emission levels was the substitution of low carbon dioxide-emitting natural gas for high-emitting coal. A weakness of earlier targets, such as those found in the Kyoto Protocol, was in its decision not to provide targets of any kind, even voluntary ones, that less-developed countries were expected to work toward. However, the second major agreement effort, the Paris Agreement, though still in its early stages, does have GHG emission targets, albeit often voluntary ones, for most countries to attempt to reach if not exceed by the year 2030. Having learned from the successes and difficulties of the Kyoto Protocol, the Paris Agreement has been more comprehensive in its desire for country-specific GHG emission-reduction targets, even if they are largely voluntary and self-selected for most developing countries.

The latter agreement has established an overall global targeted temperature increase of not more than 1.5 degrees Celsius over the temperature baseline of preindustrial temperature levels. Unfortunately, the Paris Agreement does not provide a definition for the baseline. The period 1850–1900 is often used as the preindustrial period baseline.[6] The proposed annual transfer of $100 billion from rich nations to poorer ones to fund much-needed investment in technologies to reduce GHGs could offset the desire of some of these countries to increase their use of fossil fuels to help jumpstart the growth of their economies. Although still in their early stages, the plans of many of the nations are not yet adequate to meet their stated GHG targets. Even more concerning is that preliminary performance is often not adequate to achieve the 2030 targets. Unfortunately, even if the individual emission targets are met, collectively they appear to be grossly inadequate to achieve the global targets set for 2030 and 2050 and thereby compromise the longer-term climate-control objectives. The inability of many countries to set, let alone meet, higher targets, and the fact that many of the targets of the Paris Agreement appear unlikely to be achieved translates into skepticism by a preponderance of scientific model projections about the ability of the mitigation effort to achieve the global temperature target.[7]

We now recognize the inadequacies of the current mitigation plan. First, the plan is unlikely to be fully realized, certainly not in the time frame envisioned in the Paris Agreement. The extent of future global changes and damage is fraught with uncertainty. This problem is exacerbated due to conflicts between the demand for economic development and the desire to control global warming. Controlling warming is in conflict with some of the developmental objectives of some of the world's largest and fastest-growing nations. In addition, there are financial limitations that may not be overcome by the transfer of wealth, which is a significant element of the Paris Agreement. However, an advantage of developing countries is that they are in a position, at least in principle, to skip the costs involved with early adoption of more fossil fuel facilities and leap directly to renewable energy sources.

As I argue in this volume, the implication of an analysis of the various climate forecasts and projections is that even if the targets of the Paris Agreement are met, a full transition to renewable energy is unlikely to be completed in a time frame necessary to avoid most of the massive predicted climate change damage. Rather than try to generate additional projections, I utilize estimates, or better consensus trends, directions, and magnitudes from among the experts who are projecting or forecasting these phenomena. I then compare the apparent increases in atmospheric GHG emission levels against the experts' projections of the extent to which GHGs are being mitigated (see Chapter 3).

Without question I find that the mitigation rates, both actual and projected, are judged by the experts as not nearly sufficient to offset the anticipated GHGs and temperature increases. Further, I compare the experts' assessments of societies' current and likely future success in controlling GHGs and those experts' views of the likely extent of climate-related damage in the near and medium terms (see Chapter 3). The conclusion is that, given current and likely rates of future emissions and the extent of mitigation, current projection models almost always indicate that the rate of increase of global temperature will substantially exceed the targeted maximums called for in the Paris Agreement. Exceeding those target maximums is associated with large and disruptive disturbances and damage to society.

If the common assessments of our collective inability to stop dramatic warming are correct, the global community is facing a crisis. However, society may still have a couple of decades to address the problem more fully.

TOWARD A SOLUTION?

I have proposed that the global community should add adaptation, or what I call Plan B, as a strong additional component to its current mitigation (or Plan A) approach. I have identified major sectors likely to be susceptible to extreme climate-related damage, and the discussion has explored some approaches that can be used for avoiding and providing for damage control management in each area. The current approach of collectively attempting to eliminate or severely reduce human-created GHGs is proving to be insufficient.

I have also examined the damage created by the loss of agricultural productivity, forest disturbance and ecological damage, and extreme events. However, my most intensive focus is on sea-level rise and its associated destruction during the twenty-first century. I view rising sea levels as my prime indicator of climate warming, since they seem most easily measured and monitored.[8] More importantly for damage response and adaptations, sea-level increases are most likely to generate identifiable widespread damage and hence provide among the greatest challenges to a successful focused damage adaptation response. I believe the examples provided in this volume identify many of the more important issues and types of damage related to global warming.

Despite some successes in marshaling GHG emission reduction efforts, most analysts find that the current approach appears to be inadequate to allow us to meet the temperature targets as called for by the Paris Agreement. Although some developed countries have stabilized emission levels, most of the world's countries still have high and growing levels of global GHG emissions. The problem is particularly severe in large developing countries that are beginning to experience rapid economic growth. Stabilization of emissions will be most difficult in the near future. Most analysts maintain that it is very unlikely that we can hold to even a 2 degree Celsius limit based on progress so far.[9] The problems are both technical and political. Politically, the replacement of existing fossil fuel facilities with renewable systems is not only expensive but draws resources from other development priorities. Furthermore, overall budgets are limited. This is the rationale with which the Paris Agreement called for wealthy countries offer assistance to their developing neighbors. Technically, issues also arise as to a country's balance between energy systems that are dependent on the weather, as renewable systems are, and traditional energy systems using fossil fuels or nuclear energy, which allow for production to proceed independent of the inherent and uncontrollable variability that renewables pose.

IS GEOENGINEERING THE ANSWER?

As discussed throughout this volume, adaptation involves one of two mutually compatible types of responses. First is the recognition that some types of damage associated with warming are likely to occur. The response is to try as best one can to anticipate the event and its likely damage and undertake programs and activities to minimize the destruction and its associated negative impact. A second approach associated with adaptation is to try to neutralize the warming effects of the GHGs. As we have seen, a variety of geo-engineering techniques have that potential. The tools that might be part of such an adaptation strategy have been listed and discussed at some length. One set of geoengineering responses involve earth or atmospheric reflective technologies. These would include modifications of the earth's surface and/or the planet's atmosphere in order to increase its heat-shielding function. Examples of attractive means to offset the warming effects of GHGs have been discussed, including atmospheric cooling techniques such as whitening clouds or the use of aerosols, which mimic the cooling effects of volcanic emissions. Some have argued that an approach combining both aerosol releases and cloud thinning could be effective.[10] A geoengineering approach may become inevitable as the failures of mitigation become more apparent.[11] Furthermore, an estimate by the Royal Society in 2009 found that the aerosol approach might be a thousand times cheaper than mitigation.[12]

There are other geoengineering approaches that would capture GHGs after they have been released from fossil fuels but before they reach the atmosphere, as with Carbon Capture and Storage, or techniques to capture GHGs while in the atmosphere, as with the creation of new carbon sequestrating forests. The geoengineering approaches, however, are different from the earlier adaptation responses in that they do not focus on avoiding particular localized damage, but rather they try to address the entire warming problem by reducing the entire Earth's ability to capture solar energy. Thus the focus would be on the global warming externality as a whole rather than simply focusing on a single location or specific type of damage.

For Plan B to be implemented, a new climate treaty need not be required. For most of these responses, an individual country or small group of countries could undertake most adaptive steps without necessarily receiving permission or even consulting the other signatories to the Paris Agreement.[13] The total costs of such an effort, while substantial, are still likely to be modest compared to the anticipated costs of a Plan A-type mitigation effort. It is possible that a large, wealthy country could self-finance such an effort.

As has been suggested, an advantage of an adaptation approach it that it offers the opportunity to undertake remedial activities independently of the broader collective whole. As the examples offered reveal, most of the activities are local. Protective efforts against rising sea levels, for example, although initiated to counter a global phenomenon, are localized and specific to each locality's situation. For example, in one case a seawall may be required, while another might need improved drainage facilities.

For some types of geoengineering activities, such as those intended to increase atmospheric reflectivity, some type of new global agreement may be needed. Any such activity likely to have global consequences would need an international agreement, the details of which should probably be worked out through the membership of the UNFCCC.

As emphasized throughout this volume, Plan B could be looked upon as a backup insurance policy. We have discussed in general the extent to which adaptation efforts are currently underway. They are most often stimulated by the effects of past climate events, such as hurricanes. However, anticipation often does not include consider-ations of future climate change-driven storm events or their likeli-hood of occurring.

ANOTHER POSSIBLE SOLUTION: IS IT A BRIDGE?

A rather remarkable occurrence is currently underway that has major implications for warming: the onset of natural gas as an increas-ingly important fossil fuel. Expanding exploration has resulted in the world's reserves of natural gas increasing substantially.[14] Coinciden-tally, along with the carbon dioxide and the global warming crises has been the development of hydrofracturing (fracking)[15] as a new natural gas extractive technology. The approach involves extracting natural gas, as well as oil, from natural rock materials found deep in the earth. The gas and oil hydrocarbons have long been recog-nized as resident in the rock, but the extraction process was prohibi-tively expensive. However, the newly developed technique involves cracking the rock, usually is several thousand feet below the earth's surface, and inserting hot water with sand grains under pressure to liberate the natural gas and oil and bring them to the earth's surface.

Natural gas has two useful characteristics. First, under large-scale production conditions, it is quite cheap, less costly as an energy source than most coal. Second, natural gas releases only one-half the carbon dioxide per unit of energy as coal. Therefore, if the coal power energy of the country can be replaced by natural gas energy, the carbon dioxide emissions are immediately reduced by one-half, since the same amount of energy is being created. Large portions of the US electrical power generating sector have already converted from coal to natural gas. This is one of the reasons that the United States has been so successful in reducing its GHG emissions,[16] despite not formally participating in most of the targeted GHG control pro-grams of the UN.[17]

Given the continued availability of low-cost natural gas, the 35 percent of the US power industry still using coal can be expected to convert to natural gas in a relatively short time, and with it the United States will continue to experience a substantial decrease in overall GHG emissions for a significant time period, especially since the prospects for natural gas are potentially huge. The continuing conversion to renewables will further this transition to emitting less and less GHGs.

Natural gas is also found in great abundance in Russia, Iran, and Qatar. So far, the major users have been developed countries, but other prospects are available, and shifts to natural gas need not be confined to developed countries' energy systems. For example, here in the United States, Alaska is considering an over $43 billion natural gas project oriented to supplying gas to nations in Asia.[18] Shale deposits are found widely around the world. For example,[19] Australia has large gas deposits and is well situated to export to the Asian market. The South American continent also appears to have significant natural gas resources but would need to develop the necessary infrastructure before making efficient use of its reserves.

Thus, while some markets for natural gas are rapidly replacing coal, overall the global transition is potentially massive. Whether this newer fuel source can provide a functioning bridge to a lower carbon, post-fossil-fuel world in a timely fashion remains to the seen. Natural gas obviously has a significant role to play in controlling global GHGs. However, in the long term it is only a part of a more environmentally friendly energy transition. The question is: a transition to what?

CONCLUSIONS

I have argued throughout this volume that although useful in addressing climate warming, most analysts believe that mitigation is inadequate. This view has been reinforced by the recent report of the IPCC finding that it would take a massive global effort, far more aggressive than any we have seen, to keep warming in line with the 1.5 degree Celsius target. They find that the earth is already en route to a 3 degree Celsius outcome.[20] The problem of GHG warming is overwhelming mitigation efforts. An adaptive dimension needs to be—indeed must be—added to the mitigation effort. Substantial attempts at adaptation are currently being undertaken in an effort to better anticipate and prepare for future challenges brought on by the effects of global warming. Such activities need to be continued and accelerated to expand this approach—Plan B— with additional resources being spent to limit the warming risk. A key advantage of these forms of adaptation is that little international coordination is required; nations can independently do much if they choose.

The geoengineering form of adaptation will require additional research to develop operational applications of some of the approaches that appear most promising. Second, reflecting the concerns of some observers, development programs should also undertake a careful assessment of the potential risks of these emerging geoengineering technologies. We must learn as quickly as possible the likelihood of any negative after-effects. Approaches need to be stressed that are inherently benign and/or self-limiting.

Finally, this volume looks collectively at all efforts to address climate change as an insurance policy. If the human community is to survive global warming, the ideal approach should, and indeed must, involve both mitigation (Plan A) and some forms of adaptation (Plan B).

NOTES

INTRODUCTION. CLIMATE CHANGE: WHERE ARE WE NOW?

1. Nathaniel Rich, "Losing Earth," *New York Times Magazine*, August 2018.
2. T. M. Wigley, R. Richels, and J. A. Edmonds, "Economic and Environmental Choices in the Stabilization of Atmospheric CO_2 Concentrations," *Nature* (1996): 379.
3. Elizabeth Kolbert, "The Fate of Earth," *New Yorker*, October 2017.
4. D. J. Wuebbles, D. W. Fahey, K. A. Hibbard, et al., eds., "Executive Summary," in *Climate Science Special Report: Fourth National Climate Assessment*, vol. 1 (Washington, DC: US Global Change Research Program, 2017), pp. 12–34.

CHAPTER 1. AL GORE AND THE GREENHOUSE GAS THEORY: PLAN A

1. In 1896, a seminal paper by Svante Arrhenius predicted change in the surface temperature driven by changes in the level of atmospheric carbon dioxide. An early scientific observation was that of physicist John Tyndall in the 1860s, where he suggested slight changes in the atmospheric composition could bring about climatic variations. See Roland Jackson, *The Ascent of John Tyndall* (Oxford: Oxford University Press, 2018).
2. The papers of that meeting were published in 1992 as *Economic Issues in Global Climate Change*, ed. John M. Reilly and Margot Anderson (Boulder, CO: Westview Press, 1992).
3. Al Gore, *An Inconvenient Truth* (New York: Rodale, 2006).
4. S. F. Singer, C. Star, R. Revelle. "What to Do about Climate Warming: Look before You Leap," *Cosmos Club Journal* (April 1991): 28–55.
5. Robert H. Nelson, *The New Holy Wars, Economics Religion versus Environmental Religion in Contemporary America* (University Park, PA: Penn State University Press, 2010).
6. Dana Hunter. "The Astonishing Climate-Changing Power of Plate Tectonics," *Scientific American*, December 2016. Tectonic plates move over time and could generate changes in sea currents.
7. "Public Opinion on Climate Change," Gallup Poll, Washington, DC, May 9, 2018.
8. Ibid.

9. See instructions of the IPCC as to their institutional objectives, "About the IPCC," Intergovernmental Panel on Climate Change, https://www.ipcc.ch/about/.

10. An early scientific observation was that of physicist John Tyndall in the 1860s, where he suggested that slight changes in the atmospheric composition could bring about climatic variations. See Jackson, *Ascent of John Tyndall.* Also in 1896, a seminal paper by Svante Arrhenius predicted change in the surface temperature driven by changes in the level of atmospheric carbon dioxide.

11. NASA Goddard Institute for Space Studies, last updated December 10, 2018, https://www.giss.nasa.gov/.

12. See Figure 2.2

13. ARTiFactor, "Measuring Ocean Temperature," *Science Buzz* (blog), June 25, 2010, http://www.sciencebuzz.org/blog/measuring-ocean-temperature.

14. Isotopes are the same element with slightly different characteristics within their atomic configuration. From these differences, crude estimates of temperature variation can be made.

15. See John Cook, "What Has Global Warming Done since 1998?" Skeptical Science, 2018, https://skepticalscience.com/global-warming-stopped-in-1998 -intermediate.htm, for an argument critical of temperature stabilization, even for a short period.

16. IPCC, *Climate Change 2014 Synthesis Report*, ed. Rajendra K. Pachauri, Leo Meyer, et al. (Geneva, Switzerland: Intergovernmental Panel on Climate Change, 2015), p. 48, https://www.ipcc.ch/site/assets/uploads/2018/02/SYR_AR5_FINAL_full.pdf.

17. "Ice Age/ Interglacial Cycle," video, 0:31, Koshland Science Museum, https://www.koshland-science-museum.org/multimedia/video/ice-age-interglacial-cycle.

18. Milutin Milankovitch, *The Milankovitch Theory*, 1938.

19. Fred Singer and Dennis Avery, *Unstoppable Hot Air: Global Warming Every 1500 Years* (New York: Roman & Littlefield, 2006).

20. Matthias Mengel et al., "Committee Sea Level Rise Under the Paris Legacy," *Journal of Nature Communications*, Potsdam Institute, February 20, 2018.

CHAPTER 2. NATURAL CLIMATE CHANGE: GHGs ARE NOT THE WHOLE ANSWER

1. See Will Steffen et al., "Trajectories of the Earth System in the Anthropocene," *Proceedings of the National Academy of Science*, August 6, 2018. These periods of warming and cooling are continually being reassessed. Concerns of periodic "hothouse" situations are being examined.

2. Bent Hansen, "History of the Earth's Climate," Dalum Hjallese Debat Club, August 2012, http://www.dandebat.dk/. Note extensive bibliography.

3. Albert Berger, "CO_2 and Astronomical Forcing of the Late Quaternary," *Proceedings of the First Solar and Space Weather Euroconference* (ESA Publications, 2000).

4. The figures in this chapter are taken from graphs found on NASA, data from NOAA, available online at, "Carbon Dioxide," NASA Global Climate Change, latest measurement October 2018, https://climate.nasa.gov/vital-signs/carbon-dioxide/.

5. Allison N. P. Stevens, "Factors Affecting Global Climate," *Nature Education Knowledge* 3, no. 10 (2011): 18.

6. Berger, "CO_2 and Astronomical Forcing."

7. "Continuum Density," NASA, June 14, 1994, last updated April 1, 1998, https://image.gsfc.nasa.gov/poetry/workbook/sunspot.html.

8. Sara Zielinski, "Did Climate Change Make the Norse Disappear from Greenland?" Smithsonian, December 4, 2015, https://www.smithsonianmag.com/science -nature/did-climate-change-make-norse-disappear-greenland-180957454/#J5G ysTyo8c2VTY0V.99.

9. Donald J. Easterbrook, *Evidence-Based Climate Science* (Amsterdam: Elsevier, 2016).

10. Brian Fagan, *The Great Warming: Climate Change and the Rise and Fall of Civilizations* (New York: Bloomsbury, 2010).

11. Michael E. Mann et al., "Northern Hemisphere Temperatures During the Past Millennium," *American Geophysical Union* 26, no. 6 (March 15, 1999): 759–62.

12. Nora Schultz, "Natural Mechanism for Medieval Warming Discovered," *New Scientist*, April 2, 2009, https://www.newscientist.com/article/dn16892 -natural-mechanism-for-medieval-warming-discovered/.

13. Cory Leahy, "Medieval Warm Period Not So Random," UT News, University of Texas at Austin, November 11, 2010; Yair Rosenthal et al., "Pacific Ocean Heat Content During the Past 10,000 Years," *Science* 342, no. 6158 (November 1, 2013): 617–21.

14. See Figure 2.2, "The Industrial Revolution Has Caused a Dramatic Rise in CO_2."

15. See Donald J. Easterbrook, "Medieval Warm Period," in *Evidence-Based Climate Science* (Amsterdam: Elsevier, 2016).

16. Valerie Trouet et al., "Latitudinal Gradients of Tree-Ring Carbon and Oxygen Isotopes," *Journal of Geophysical Research* 121, no. 7 (July 2016): 1978–91.

17. See Easterbrook, "Medieval Warm Period," for evidence that the Medieval Warming was warmer and involved an extensive area.

18. Ibid.

19. Warmer weather allows trees to grow higher up the hill and mountainsides. Even after the trees have died due to cooler weather the vestiges of the highest tree line can remain for centuries thus providing evidence of an earlier warmer period. Also see Leif Kullman, "Ecological Tree Line History and Palaeoclimate," *Boreas* 42, no. 3 (January 2, 2013): 555–67; Keith Sherwood and Craig Idso, "The Broad View of Holocene Climate from the Swedish Scandes," *CO_2 Science* 16, no. 50 (December 11, 2013).

20. Thomas G. Moore, "Happiness Is a Warm Planet," *Hoover Digest* (Hoover Institute), no. 1 (January 30, 1998). Also H. H. Lamb, *Climate History and the Modern World* (New York: Rutledge, 1982).

21. Wallace S. Broecker, *The Great Ocean Conveyer* (Princeton, NJ: Princeton University Press, 2011).

22. Fagan, *Great Warming*.

23. David Archer, *The Global Carbon Cycle* (Princeton, NJ: Princeton University Press, 2010).

24. See Figure 2.2.

25. Werner Schmutz et al., "Sun's Impact on Climate Change Quantified for the

First Time," Swiss National Science Foundation, March 27, 2017, http://www.snf.ch/en/ researchinFocus/newsroom/Pages/news-170327-press-release-suns-impact-on-climate -change-quantified-for-first-time.aspx.

26. See Archer, *Global Carbon Cycle*; also figure 2.1.

27. Archer, *Global Carbon Cycle*.

28. Tom Wigley and Ben Santer, "A Probabilistic Quantification of the Anthropogenic Component of Twentieth Century Global Warming," *Climate Dynamics* 40, no. 5–6 (March 2013): 1087–1102, in examination of the 2007 IPCC Fourth Assessment Report.

29. IPCC, *Climate Change 2014 Synthesis Report*, ed. Rajendra K. Pachauri, Leo Meyer, et al. (Geneva, Switzerland: Intergovernmental Panel on Climate Change, 2015), p. 48, https://www.ipcc.ch/site/assets/uploads/2018/02/SYR_AR5_FINAL_full.pdf.

30. Henrik Svensmark et al., "Increased Ionization Supports Growth of Aerosols into Cloud Condensation Nuclei," *Nature Communications*, 8, no. 2199 (2017).

31. Angela Olinto, "NASA's New Probe," *Wall Street Journal*, August 10, 2018, p. A13.

32. *Encyclopaedia Britannica*, s.v. "Maunder Minimum," October 16, 2016, https:// www.britannica.com/science/Maunder-minimum.

33. Schmutz, et al., "Sun's Impact on Climate Change."

34. Ibid.

35. Ibid.

36. Ibid.

37. Sam White, *A Cold Welcome: The Little Ice Age and Europe's Encounter with North America* (Harvard, MA: Harvard University Press, 2017).

38. Archer, *Global Carbon Cycle*.

39. David Archer, *Global Warming: Understanding the Forecast*, 2nd ed. (Hoboken, NJ: Wiley, 2011), chap. 10.

40. Ibid.

41. "Episode 13: Kelvin Droegemeier Talks about the Past, Present and Future of Weather Prediction," Kelvin Droegemeier, National Science/Policy Landscape, STEM Talk, June 8, 2016, YouTube video, 1:09:36.

42. B. David A. Naafs et al., "High Temperatures in the Terrestrial Mid-Latitudes during the Early Palaeogene," *Natural Geoscience* 11 (July 30, 2018): 766–71. Climate models poorly predicted the ancient past. For example, the temperature associated with a warming of the Eocene Epoch of three million years ago requires a much higher level of CO_2 than climate current models suggest. Inability of climate models to accurately predict past events suggests we view climate model future projections with great care.

43. R. D. Cess et al., "Uncertainties in Carbon Dioxide Radiative Forcing in Atmospheric General Circulation Models," *Science* 262, no. 5137 (1993): 1252–55.

44. William Collins et al., "Radiative Forcing by Well-Mixed Greenhouse Gases: Estimates from Climate Models in the Intergovernmental Panel on Climate Change (IPCC) Fourth Assessment Report (AR4)," *Journal of Geophysical Research* 111, no. D14 (July 28, 2006).

45. B. J. Soden et al., "Reducing Uncertainties in Climate Models," *Science* 361, no. 6400 (July 27, 2018): 326–27.

46. See particularly Schmutz, et al., "Sun's Impact on Climate Change."

47. Ibid.

48. Ibid.

49. Florian Adolphi, Raimund Muscheler et al., "Persistent Link between Solar Activity and Greenland Climate during the Last Glacial Maximum," *Nature Geoscience* 7, no. 9 (2014): 662–66.

50. Svensmark et al., "Increased Ionization Supports Growth."

51. See Chapter 5.

52. "El Niño & La Niña (El Niño-Southern Oscillation)," NOAA Climate.gov, November 2018, https://www.climate.gov/enso.

53. Figure 2.2 data provides evidence of a stable carbon level. Archer, *Global Warming*, provides a hypothetical explanation with his discussions of ocean cycles and climate change. The hypothesis that these are driven by variations in solar intensity has been suggested by some of the studies cited earlier. However, this hypothesis is not confirmed. This book argues for more climate change preventive activity on the supposition that the human-driven levels are substantial in themselves and may possibly be supplemented by nature warming from whatever source, perhaps solar.

54. Gen. 46: 5–8.

55. Eric H. Cline, *1177 B.C.: The Year Civilization Collapsed* (Princeton, NJ: Princeton University Press, 2015).

56. Kyle Harper, *The Fate of Rome: Climate, Disease and the End of an Empire* (Princeton, NJ: Princeton University Press, 2017).

57. Michael McCormick et al., "Climate Change During and After the Roman Empire," *Journal of Interdisciplinary History* 43, no. 2 (August 2012): 169–220.

58. *Wikipedia*, s.v. "Medieval Warm Period," last edited October 2, 2018, https://en.wikipedia.org/wiki/Medieval_Warm_Period.

59. Schultz, "Natural Mechanism for Medieval Warming."

60. Archer, *Global Warming*.

61. *Encyclopaedia Britannica*, s.v. "Maunder Minimum."

62. Sam White, *A Cold Welcome: The Little Ice Age and Europe's Encounter with North America* (Harvard, MA: Harvard University Press, 2017).

63. Having lived some of my life in Delaware, I can assure the reader that serious ice no longer flows out of the Delaware River.

64. Brian M. Fagan, *The Little Ice Age: How Climate Made History, 1300–1850* (New York: Basic Books, 2000).

65. Ibid.

66. Wallace S. Broecker, *The Great Ocean Conveyer* (Princeton, NJ: Princeton University Press, 2011).

67. See graph from NASA, data from NOAA, available online at, "Proxy Indirect Measurements," NASA Global Climate Change, latest measurement October 2018, https://climate.nasa.gov/vital-signs/carbon-dioxide/.

68. Svensmark et al., "Increased Ionization Supports Growth."

69. "More Americans Accepted Global Warming during Hot Spring," Climatewire, July 11, 2018.

CHAPTER 3. PLAN A: MITIGATION— A BRIDGE TOO FAR?

1. Arianna Skibell, "World Unlikely to Meet Temperature Goal—Leaked U.N. Report," *Greenwire*, February 15, 2018, https://www.eenews.net/greenwire/stories/ 1060074021?t=https%3A%2F%2Fwww.eenews.net%2Fstories%2F1060074021. Draft report being developed in the IPCC as their Summary for Policy Makers (SPM) for the forthcoming Sixth Assessment Report, due in 2019.

2. David Hone, "Meeting the Goals of the Paris Agreement," *Shell Oil Climate Change* (blog), March 27, 2018, https://blogs.shell.com/2018/03/27/meeting-the -goals-of-the-paris-agreement/.

3. Mark Dwortzan, "How Much of a Difference Will the Paris Agreement Make?" MIT News, April 22, 2016, http://news.mit.edu/2016/how-much -difference-will-paris-agreement-make-0422.

4. David Wallace-Wells, "'Personally, I Would Rate the Likelihood of Staying Under Two Degrees of Warming as Under 10 Percent': Michael Oppenheimer on the 'Unknown Unknowns' of Climate Change," *New York Magazine*, July 13, 2017, http:// nymag.com/intelligencer/2017/07/michael-oppenheimer-10-percent-chance-we-meet -paris-targets.html.

5. Alex Lenferna, "The Costs of Solar Geoengineering," Ethics & International Affairs, July 2017, https://www.ethicsandinternationalaffairs.org/2017/costs-of -geoengineering/.

6. Massachusetts v. Environmental Protection Agency 549 U.S. 497 (2007).

7. For example see: Warwick J. McKibbin et al., "The Potential Role of a Carbon Tax in U.S. Fiscal Reform," Brookings, July 24, 2012, https://www.brookings.edu/ research/the-potential-role-of-a-carbon-tax-in-u-s-fiscal-reform/.

8. *20 Years of Carbon Capture and Storage: Accelerating Future Deployment* (Paris: International Energy Agency, November 2016).

9. *Wikipedia*, s.v. "Carbon Emissions Trading," last edited November 19, 2018, https://en.wikipedia.org/wiki/Carbon_emission_trading.

10. John C. Stringer, "Opportunities for Carbon Control in the Electric Power Industry," in *Carbon Management: Implications for R&D in the Chemical Sciences and Technology* (Washington, DC: National Academies Press, 2001), chap. 4.

11. Roger Sedjo et al., *The Economics of Carbon Sequestration in Forestry* (Boca Raton, FL: CRC Press, 1997).

12. *20 Years of Carbon Capture and Storage.*

13. Swiss Federal Laboratories, "Batteries of the Future: Low-Cost Battery from Waste Graphite," *ScienceDaily*, October 11, 2017.

14. "Bioenergy," US Department of Energy, https://www.energy.gov/science -innovation/energy-sources/renewable-energy/bioenergy.

15. The concept in forestry of a regulated forest is one where net harvest is offset by an equivalent net growth, with no change in the volume of wood in the forest. Since volume is directly related to carbon, a regulated forest releases no net carbon.

16. *Future of America's Forests and Rangelands: Update to the Forest Service 2010 Resources Planning Act Assessment* (2010; Washington, DC: US Department of Agriculture, September 2016).

17. Ibid.

18. Michael Chavez, "The Kyoto Protocol: Accomplishments and Failures" (paper, Department of Atmospheric Sciences, College of the Environment, University of Washington, Seattle, 2009).

19. Hydraulic fracturing (fracking); a system to extract oil and natural gas by fracturing rock with a pressurized liquid.

20. Natural gas releases roughly one-half the amount of carbon dioxide from natural gas per unit of energy compared to coal.

21. Climate Action Tracker, 2018, https://climateactiontracker.org/.

22. "Global Greenhouse Gas Emissions Data," US Environmental Protection Agency, last updated April 13, 2017, https://www.epa.gov/ghgemissions/global-greenhouse-gas-emissions-data.

23. "The Paris Agreement," United Nations Climate Change, last updated October 22, 2018, https://unfccc.int/process-and-meetings/the-paris-agreement/the-paris-agreement.

24. Ibid.

25. Liz Peek, "China's Rising Emissions Prove Trump Right on Paris Agreement," *Hill*, June 5, 2018, https://thehill.com/opinion/energy-environment/390741-chinas-rising-emissions-prove-trump-right-on-paris-agreement.

26. "China," Climate Action Tracker, last updated November 30, 2018, https://climateactiontracker.org/countries/china/fair-share/.

27. China is becoming a major gas importer from the US.

28. *Report on the Environment: Energy Use* (Washington, DC: US Environmental Protection Agency, 2018), https://www.epa.gov/roe/.

29. "Biomass: Renewable Energy from Plants and Animals," US Energy Information Administration, last updated June 21, 2018, https://www.eia.gov/energyexplained/?page=biomass_home.

30. Sören Amelang, "Germany on Track to Widely Miss 2020 Climate Target – Government," Journalism for the Energy Transition, June 13, 2018, https://www.cleanenergywire.org/news/germany-track-widely-miss-2020-climate-target-government; Paul Hockenos, "Carbon Crossroads," *Yale Environment 360*, December 13, 2018.

31. Daniel Raimi and Alan Krupnick, "Decarbonization: It Ain't That Easy," *RFF* (blog), April 20, 2018, http://www.rff.org/blog/2018/decarbonization-it-ain-t-easy.

32. Patrick McGeehan, "Cuomo Confirms Deal to Close Indian Point Nuclear Plant," *New York Times*, January 9, 2017, https://www.nytimes.com/2017/01/09/nyregion/cuomo-indian-point-nuclear-plant.html.

33. Robert Bryce, "Andrew Cuomo's Wind Farm," *Wall Street Journal*, May 19, 2018, p. A11.

34. Climate Action Tracker, https://climateactiontracker.org/.

35. Raimi and Krupnick, "Decarbonization."

36. Vanessa Schipani, "Will Paris Have a 'Tiny' Effect on Warming?" FactCheck.org, June 14, 2017, https://www.factcheck.org/2017/06/will-paris-tiny-effect-warming/.

37. David Roberts, "There's a Huge Gap between the Paris Climate Change Goals and Reality," *Vox*, November 6, 2017, https://www.vox.com/energy-and-environment/2017/10/31/16579844/climate-gap-unep-2017.

38. Climate Action Tracker, https://climateactiontracker.org/.

39. Climate Action Tracker, February 2, 2018, https://climateactiontracker.org/.

40. e = equivalent when adjusting for noncarbon GHGs. An adjustment to account for the carbon equivalent effect of noncarbon GHGs.

41. Susan Joy Hassol, "Questions and Answers: Emissions Reductions Needed to Stabilize Climate" (Washington, DC: Presidential Climate Action Project, 2014).

42. *Subcommittee on Environment and Subcommittee on Energy Hearing - Geoengineering: Innovation, Research, and Technology*, 115th Cong. (November 8, 2017) (statement by Committee Chairman Lamar Smith, R-Texas).

43. David Archer, *The Global Carbon Cycle* (Princeton, NJ: Princeton University Press, 2010).

44. Although funding for geoengineering in the past has been low, my impression is that interest in this approach is growing even as I have been writing this volume. I anticipate that more funds are becoming available.

45. Lenferna, "The Costs of Solar Geoengineering." Harvard's David Keith estimates the geoengineering costs at roughly $10 billion.

46. Norman J. Rosenberg et al., *Greenhouse Warming: Abatement and Adaptation* (proceedings of a workshop held in Washington, DC, June 14–15, 1988; Resources for the Future, 1989).

47. Adrian Raftery et al., "Less than 2 Degree C Warming by 2100 Unlikely," *Nature Climate Change* 7 (2017): 637–41.

CHAPTER 4. PLAN B: THE ADAPTATION SOLUTION

1. *FEMA: Prioritizing a Culture of Preparedness* 115th Cong. (April 11, 2018) (statement of FEMA Administrator Brock Long before the Senate Homeland Security Committee).

2. "The Paris Climate Agreement," United Nations, last updated October 22, 2018, https://unfccc.int/process-and-meetings/the-paris-agreement/the-paris -agreement.

3. "Climate Watch," World Resources Institute, https://www.wri.org/our-work/ project/climate-watch.

4. Ibid.

5. NASA Global Climate Change, "Carbon Dioxide," https://climate.nasa.gov/; Climate Action Tracker, https://climateactiontracker.org/.

6. The atmosphere has an ability to cleanse itself of carbon dioxide. However, this process is relatively slow and is being overwhelmed by incoming flows.

7. Gary Griggs et al., *Rising Seas in California: An Update on Sea-Level Rise Science* (Oakland, CA: Ocean Science Trust, April 2017).

8. Michael Geddes, *Making Public Private Partnerships Work: Building Relationships and Understanding Cultures* (Aldershot: Glower, 2005).

9. Arthur Charpentier, "Insurability of Climate Risks," *Geneva Papers on Risk and Insurance: Issues and Practice* 33, no. 1 (January 2008): 91–109.

10. Jonathan L. Bamber et al., "Reassessment of the Potential Sea-Level Rise from a Collapse of the West Antarctic Ice Sheet," *Science* 324, no. 5929 (May 15, 2009): 901–903.

11. Andrew Shepherd, et al., "Mass Balance of the Antarctic Ice Sheet from 1992– 2017," *Nature* 558 (2018): 219–22.

12. "Is Sea Level Rising?" National Ocean Service, last updated June 25, 2018, https://oceanservice.noaa.gov/facts/sealevel.html.

13. National Research Council, "Chapter 5: Projections of Sea-Level Change," in *Sea-Level Rise for the Coasts of California, Oregon, and Washington* (Washington, DC: National Academies Press, 2012).

14. Scott Waldman, "Lawmaker Says Tumbling Rocks Are Causing Seas to Rise," Climate Wire, May 17, 2018.

15. Fred Singer, a critical climate scientist, discussed the issue in the *Wall Street Journal*, noting the complexities of explaining sea-level rise. Fred Singer, "The Sea Is Rising, but Not Because of Climate Change," *Wall Street Journal*, May 15, 2018.

16. Andrew Shepherd et al., "Warm Ocean Is Eroding West Antarctic Ice Sheet," *Geophysical Research Letters* 31, no. 23 (June 13, 2018).

17. Interestingly, great ice formations exert a gravitational pull based on their mass. As the formation melts, its mass diminishes, and so does its gravitational force.

18. Harold R. Wanless, "South Florida's Sea-Level Threat Is Worse than You Think," *South Florida Sun-Sentinel*, June 20, 2018.

19. California is anticipating the impacts of climate change in a variety of areas. For example, a proposal to replace an old gas pipeline is meeting resistance with the argument that shifting fuel uses, due to climate change, will make the new pipeline unnecessary. Rob Nikolewski, "Ratepayer Group Says SDG&E and SoCalGas Made False Statements about Natural Gas Pipeline," *San Diego Union-Tribune*, June 20, 2018, https://www.sandiegouniontribune.com/business/energy-green/sd-fi-ora-motion-20180620-story.html.

20. Patrick W. Limber et al., "A Model Ensemble for Projecting Multidecadal Coastal Cliff Retreat During the 21st Century," *Journal of Geophysical Research: Earth's Surface* 123, no. 7 (July 2018): 1566–89.

21. Rosanna Xia, "Rising Sea Level Poses Growing Threat to Coast," *San Diego Union-Tribune*, June 27, 2018, P. A8.

22. Shepherd et al., "Warm Ocean Is Eroding."

23. *San Diego Union-Tribune*, September 6, 2017, p B5.

24. Daniel Botkin, "What's the Likely Future of New Orleans?," *Daniel B. Botkin* (blog), April 7, 2007, https://www.danielbbotkin.com/2007/04/07/whats-the-likely-future-of-new-orleans-history-tells-us-whats-likely/.

25. *Harris County Supplemental Action Plan* (Harris County, TX, submitted for public review on June 27, 2018 to July 10, 2018).

26. Dan Frosch, "Houston Voters Back $2.5 Billion Plan to Bolster Flood Defenses After Harvey," *Wall Street Journal* August 26, 2018, https://www.wsj.com/articles/houston-voters-back-2-5-billion-plan-to-bolster-flood-defenses-after-harvey-1535256942.

27. First Street Foundation, "As the Seas Have Been Rising Home Values Have Been Sinking," press release.

28. R. Epanchin-Niell et al., "Investing in Coastal Protected Lands under Threat from Sea-Level Rise," *Resources*, no. 194 (Spring 2017), http://www.rff.org/research/publications/investing-coastal-protected-lands-under-threat-sea-level-rise.

29. Ibid.

30. Wing et al., "Estimates of Present and Future Flood Risk in the Conterminous United States," *Environmental Research Letters* 13 (2018).

31. Selma Guerreiro et al., "Future Heat-Waves, Drought and Floods in 571 European Cities," *Journal of Environmental Research Letters* (January 2018).

32. Vinod Thomas, "Building Better Defenses against Rising Floods and Storms," September 5, 2017, Brookings, https://www.brookings.edu/blog/future -development/2017/09/05/building-better-defenses-against-rising-floods-and-storms/.

33. John Fialka, "World Faces Sharp Rise in Tropical Storm Damage Risk," *Scientific American*, September 20, 2017, https://www.scientificamerican.com/article/world-faces -sharp-rise-in-tropical-storm-damage-risk/?utm_source=facebook&utm_medium =social&utm_campaign=sa-editorial-social&utm_content&utm_term=sustainability _partner_.

34. The National Flood Insurance Program, Department of Homeland Security, last updated November 15, 2018, https://www.fema.gov/national-flood-insurance-program.

35. *Wikipedia*, s.v. "National Flood Insurance Program," last updated October 16, 2018, https://en.wikipedia.org/wiki/National_Flood_Insurance_Program.

36. Only twenty percent of homes in flood prone areas have federal flood insurance.

37. How much does federal flood insurance cost? The average federal discount is about sixty to sixty-six percent of the full cost of the insurance.

38. Over five million homes have flood insurance in the US. FEMA paid over $8.4 billion for damages after Sandy in 2012 and $16 billion after Katrina in 2005.

39. People of the State of California v. BP et al., San Francisco Superior Court Case no. CGC 17-561370; People of the State of California v. BP et al., Alameda County Superior Court Case no. RG17875889.

40. Roger Sedjo and Allen Solomon, "Climate and Forests," in *Greenhouse Warming: Abatement and Adaptation*, ed. Norman J. Rosenberg, William E. Easterling III, Pierre R. Crosson, and Joel Darmstadter (Washington, DC: Resources for the Future, 1989), chap. 8, pp. 105–19.

41. Obviously, the substitute product would be one where a market already exists.

42. P. Kauppi and M. Posch, "Sensitivity of Boreal Forests to Possible Climatic Warming," *Climatic Change* 7, no. 1 (1985): 45–54.

43. Nancy Harris and Michael Wolosin, "Ending Tropical Deforestation: Tropical Forests and Climate Change; The Latest Science," World Resources Institute, June 2018, https://www.wri.org/publication/ending-tropical-deforestation -tropical-forests-and-climate-change-latest-science.

44. C. Le Quere et al., "Global Carbon Budget 2017," *Earth System Science Data* 10, no. 1 (2018): 405–48, https://www.earth-syst-sci-data.net/10/405/2018/.

45. Thin Lei Win, "Mangrove-Planting Drones on a Mission to Restore Myanmar Delta," Reuters, August 21, 2017, https://www.reuters.com/article/us-myanmar -environment-mangroves-tech/mangrove-planting-drones-on-a-mission-to -restore-myanmar-delta-idUSKCN1B10EQ.

46. Brent Sohngen, Robert Mendelsohn, and Roger Sedjo, "A Global Model of Climate Change Impacts on Timber Markets," *Journal of Agricultural and Resources Economics* 26, no. 2 (2001): 326–43.

47. H. Shugart et al., *Forests and Global Change: Potential Impacts on U.S. Forest Resources* (Arlington, VA: Pew Center for Climate Change, February 2003), p. 52.

48. Sohngen, Mendelsohn, and Sedjo, "Global Model of Climate Change."

49. General Circulation Models represent changes in physical processes, in this case on the land surface, in response to a simulate increase in GHG accumulations.

50. An even-aged forest involves the vast majority of trees of the dominant species being approximately of the same age, growing, and indeed often dying together.

51. This phenomenon is similar to the birth of a litter in the animal kingdom where the later-arriving newborns are often disadvantaged in survival and out-competed by their older siblings.

52. Carbon constitutes roughly one-half the weight of a tree. That released by a newly burned tree vastly exceeds that captured by a small seedling in the near term. Obviously, as the tree grows over the years to the size of its predecessor, it will gradually capture an amount of carbon equivalent to that released initially.

53. *Wildlife in a Warming World* (Gland, Switzerland: World Wildlife Fund, March 13, 2018).

54. Brad Plumer, "Stitching Together Forests Can Help Save Species, Study Finds," *New York Times*, August 21, 2017, https://www.nytimes.com/2017/08/21/climate/rain-forest-corridors-species-habitats-extinctions.html; Josh Lew, "10 Important Wildlife Corridors," Mother Nature Network, September 4, 2015, https://www.mnn.com/earth-matters/wilderness-resources/stories/10-important-wildlife-corridors.

55. See Brent Sohngen and Robert Mendelsohn, "An Optimal Control Model of Forest Carbon Sequestration," *American Journal of Agricultural Economics* 85, no. 22003 (2003): 448–57, for an application of GCM to forestry.

56. M. Hulme et al., "Precipitation Sensitivity to Global Warming," *Geophysical Research Letters* 25, no. 7 (1998).

57. Roger Pielke Jr., "The Hurricane Lull Couldn't Last," *Wall Street Journal*, August 31, 2017, https://www.wsj.com/articles/the-hurricane-lull-couldnt-last-1504220969.

58. *Benefits of Reduced Anthropogenic Climate Change* (Boulder, CO: National Center for Atmospheric Research).

59. This phrase has been attributed to numerous individuals including, recently, the major of Chicago, Rahm Emanuel.

60. Sheri Fink, "Puerto Rico's Hurricane Maria Death Toll Could Exceed 4,000, New Study Estimates," *New York Times*, May 29, 2018, https://www.nytimes.com/2018/05/29/us/puerto-rico-deaths-hurricane.html?rref=collection%2Fbyline%2Fsheri-fink. A later report estimated the loss of life at just under 3000.

61. Arturo Massol-Deyá, Jennie C. Stephens, and Jorge L. Colón, "Renewable Energy for Puerto Rico," *Science* 362, no.6410 (October 5, 2018): 7.

62. Ibid.

63. Francine McKenna, "Congressional Hearing Debates Various Proposals to Fix Bankrupt Puerto Rico Electric Power Authority," MarketWatch, July 25, 2018, https://www.marketwatch.com/story/congressional-hearing-debates-various-proposals-to-fix-bankrupt-puerto-rico-electric-power-authority-2018-07-25.

64. *Full Committee Hearing to Examine Puerto Rico's Electric Grid*, U.S. Senate Committee on Energy and Natural Resources (May 8, 2018).

65. Kathya Severino Pietri, "Disaster Aid for Puerto Rico: Congressional Actions," Centro for Puerto Rican Studies, https://centropr.hunter.cuny.edu/events-news/rebuild-puerto-rico/policy/disaster-aid-puerto-rico-congressional-actions.

66. Luke Richardson, "New Solar Panels: What's Coming to Market in 2018?" Energy Sage, April 19, 2018, https://news.energysage.com/new-solar-panels-whats-coming-market-2018/.

67. Massol-Deyá, Stephens, and Colón, "Renewable Energy for Puerto Rico."

68. Department of Energy, "More Work to Be Done to Strengthen Puerto Rico's Electrical Grid," press release, May 9, 2018

69. Jamie Condliffe, "Can Tesla Reboot Puerto Rico's Power Grid?" *MIT Technology Review*, October 6, 2017.

70. Adam Rodgers, "Why Can't We Fix Puerto Rico's Electrical Power Grid?" *Wired*, April 18, 2018, https://www.wired.com/story/why-cant-we-fix-puerto-ricos-power-grid/.

71. Nicole Acevedo, "Puerto Rico's New Law Moves to Privatize Power Grid Nine Months after Hurricane Maria," NBC News, June 20, 2018, https://www.nbcnews.com/storyline/puerto-rico-crisis/puerto-rico-officially-moves-privatize-power-grid-9-months-after-n885111.

72. The Jones Act limited the ships used for inter-US transport to those built in US shipyards.

73. American-built transport ships are relatively few and expensive since they require American crews.

74. Matt Wirz, "Why Puerto Rico Is Proving to Be 2018's Top Bond Investment," *Wall Street Journal*, March 18, 2018, https://www.wsj.com/articles/puerto-rico-bonds-are-a-surprise-star-performer-as-economy-starts-to-mend-1521115200.

75. Ike Brannon, "Simply Forgiving Puerto Rico's Debt Would Be a Mistake," CNBC, October 12, 2017, https://www.cnbc.com/2017/10/12/simply-forgiving-puerto-ricos-debt-would-be-a-huge-mistake-commentary.html.

76. *Climate Change Fourth Assessment Report*, vol. 3 (Geneva, Switzerland: United Nations Intergovernmental Panel on Climate Change, 2007).

77. Herika Kummel and Randy Jackson, "Potential Carbon Sequestration with C4 Grasses Abundance in Restored Prairie of Southern Wisconsin," in *Wisconsin Integrated Crop Trial, Twelfth Report: 2007 & 2008*, ed. Janet Hedtcke and Josh Posner (Madison: University of Wisconsin College of Agricultural Arts and Sciences, May 2010).

78. C4 plants make four-carbon sugar instead of the two three-carbon sugars of C3 plants.

79. Rosenberg, Easterling, Crosson, and Darmstadter, ed., *Greenhouse Warming*.

80. "The United States Uses a Mix of Energy Sources," US Energy Information Administration, last updated May 16, 2018, https://www.eia.gov/energyexplained/index.php?page=us_energy_home; "Biomass: Renewable Energy from Plants and Animals," US Energy Information Administration, last updated June 21, 2018, https://www.eia.gov/energyexplained/?page=biomass_home.

81. "Biomass: Renewable Energy."

82. Sedjo and Solomon, "Climate and Forests."

83. "Cap and Trade Basics," Center for Climate and Energy Solutions, https://www.c2es.org/content/cap-and-trade-basics/.

84. Issues exist with fracking particularly with respect to water use and purity.

85. Glenn Kessler, "John Kerry's Misfire on U.S. Performance on Kyoto Emissions Targets," *Washington Post, Fact Checker* (blog) May 30, 2013.

86. James Temple, "The Carbon-Capture Era May Finally Be Starting," *MIT Technology Review*, February 20, 2018, https://www.technologyreview.com/s/610296/the-carbon-capture-era-may-finally-be-starting/.

87. David L. Chandler, "New Battery Gobbles Up Carbon Dioxide," *MIT News*, September 21, 2018, http://news.mit.edu/2018/new-lithium-battery-convert-carbon-dioxide.

88. Bronte Lord, "This Concrete (Yes, Concrete) Is Going High-Tech," CNN, July 6, 2018, http://money.cnn.com/2018/06/12/technology/concrete-carboncure/index.html.

89. Warren Cornwall, "Cement Soaks Up Greenhouse Gases," *Science*, November 21, 2016, https://www.sciencemag.org/news/2016/11/cement-soaks-greenhouse-gases.

90. *Subcommittee on Environment and Subcommittee on Energy Hearing - Geoengineering: Innovation, Research, and Technology*, 115th Cong. (November 8, 2017) (statement by Committee Chairman Lamar Smith, R-Texas).

CHAPTER 5. ADAPTATION THROUGH REFLECTIVITY AND GEOENGINEERING

1. "How Could Expanding Gas Cause Heat Loss?" Qualitative Reasoning Group, Northwestern University, http://www.qrg.northwestern.edu/projects/vss/docs/thermal/3-how-could-expanding-gas-cause-heat-loss.html.

2. Clean Air Act of 1963, 42 U.S.C. § 7401 (1990).

3. *Wikipedia*, s.v. "Climate Engineering," last updated November 11, 2018, https://en.wikipedia.org/wiki/Climate_engineering.

4. Naomi Klein, *This Changes Everything: Capitalism vs. The Climate* (London: Penguin, 2015).

5. David Archer, *The Global Carbon Cycle* (Princeton, NJ: Princeton University Press, 2010). The book discusses the flow of carbon among the atmosphere, ocean, soils, and mineral rocks.

6. P. J. Crutzen, "Albedo Enhancement by Stratospheric Sulfur Injections," *Climatic Change* 77 (2006): 211–19.

7. J. Templeton, "The Daunting Math of Climate Change Means We Will Need Carbon Capture," *MIT Technology Review*, April 24, 2018, https://www.technologyreview.com/s/610927/the-daunting-math-of-climate-change-means-well-need-carbon-capture/.

8. Naomi Klein, "Geoengineering: Testing the Waters," *New York Times*, October 27, 2012.

9. J. Proctor et al., "Estimating Global Agricultural Effects of Geoengineering," *Nature* 560 (August 8, 2018): 480–83.

10. David Victor et al., "The Geoengineering Option: A Last Resort against Global Warming?" *Foreign Affairs*, March/April 2009.

11. Scott Barrett, "The Incredible Economics of Geoengineering," *Environmental and Resource Economics* 39, no. 1 (2008): 45–54.

12. For example, some geoengineering approaches involve periodic sprinkling of the atmosphere. James Burgess, "Geoengineering: Sprinkling Mineral Dust in Oceans to Combat Climate Change," OilPrice.com, January 22, 2013, https://oilprice.com/Latest-Energy-News/World-News/Geoengineering-Sprinkling-Mineral-Dust-in-Oceans-to-Combat-Climate-Change.html.

13. Albedo: the amount of energy reflected back from the earth's surface. The range is from zero, on reflection, to one.

14. Note, however, that a forest that maintains a snow cover has a high reflectivity compared to a forest without snow.

15. Ken Buesseler, "Fertilizing the Ocean with Iron," in *Woods Hole Oceanographic Institute 1999 Annual Report* (Woods Hole, MA: WHOI, 1999).

16. Geraint Tarling and Sally E. Thorpe, "Oceanic Swarms of Antarctic Krill Perform Satiation Sinking," *Proceedings of the Royal Society B* 284, no. 1869 (December 13, 2017).

17. Andrea Thompson, "Krill Are Disappearing from Antarctic Waters," *Scientific American*, August 29, 2016, https://www.scientificamerican.com/article/krill -are-disappearing-from-antarctic-waters/.

18. Roger Sedjo et al., *Economics of Carbon Sequestration Forestry* (Boca Raton, FL: CRC Press, 1997).

19. Forest wildfires have been part of the natural ecosystem for eons. Wildfires, which are started by natural processes such as lighting strikes, have cleaned and recycled the materials of dead forests for ages without the help of humans. Thus, over time, the natural system probably recycles similar amounts of material and thus normally does not affect the level of atmospheric GHGs much. However, human activities can change this balance.

20. "Chapter 4," in *Climate Change Third Assessment Report* (Geneva, Switzerland: United Nations Intergovernmental Panel on Climate Change, 2001), p. 201.

21. Ibid., pp. 334–36.

22. Technically the title was co-convening lead author.

23. *Subcommittee on Environment and Subcommittee on Energy Hearing - Geoengineering: Innovation, Research, and Technology*, 115th Cong. (November 8, 2017) (statement by Committee Chairman Lamar Smith, R-Texas).

24. Representative Lamar Smith was Chairman of the House Science Committee before his retirement in 2018.

25. Klein, *This Changes Everything*.

26. Scott Barrett, "Solar Engineering's Brave New World," *Review of Environmental Economics and Policy* 8, no. 2 (July 11, 2014).

27. See Gernot Wagner and Martin L. Weitzman, *Climate Shock* (Princeton, NJ: Princeton Press, 2015), for a discussion of the free rider in climate change.

28. For details see Harvard's Solar Geoengineering Research Program, Harvard University Center for the Environment, 2018, https://geoengineering.environment. harvard.edu/.

29. University of Washington Joint institute for Atmosphere and Oceans, 2018, https://jisao.uw.edu/.

30. *Subcommittee on Environment and Subcommittee on Energy Hearing*.

31. Carlo R. Carere, "Mixotrophy Drives Niche Expansion of Verrucomicrobial Methanotrophs," *ISME Journal* 11 (2017): 2599–610.

32. John Fialka, "Experts Grapple with Side Effects of Shading Earth," Climatewire, July 5, 2018, https://www.eenews.net/climatewire/2018/07/05/stories/1060087893.

33. John Fialka, "The Best Way to Shade Earth," *Scientific American*, July 5, 2018, https://www.scientificamerican.com/article/the-best-way-to-shade-earth/.

34. "Clouds: The Wild Card in Climate Change" (Alexandria, VA: National Science Foundation).

35. Andy Jones et al., "Climate Impacts of Geoengineering Marine Stratocumulus Clouds," *Journal of Geophysical Research* 114, no. D10 (May 27, 2009).

36. "Norway: StatoilHydro's Sleipner Carbon Capture and Storage Project Proceeding Successfully," Energy-pedia News, March 8, 2009, https://www .energy-pedia.com/news/norway/statoilhydros-sleipner-carbon-capture-and-storage -project-proceeding-successfully.

37. D. Keighley and C. Maher "A Preliminary Assessment of Carbon Storage Suitability in Deep Underground Geological Formations of New Brunswick," *Atlantic Geology* 51 (2015): 269–86.

38. Emma Merchant, "Can Updated Tax Credits Bring Carbon Capture Into the Mainstream?" Green Tech Media, February 22, 2018, https://www.greentechmedia.com/articles/read/can-updated-tax-credits-make-carbon-capture-mainstream#gs.ofGN60Q.

39. Umar Irfan, "Will Carbon Capture and Storage Ever Work?" *Scientific American*, May 25 2017, https://www.scientificamerican.com/article/will-carbon-capture-and-storage-ever-work/.

40. Sarang Supekar and Steve Skerlos, "The Latest Bad News on Carbon Capture from Coal Power Plants: Higher Costs," The Conversation, December 3, 2015, https://theconversation.com/the-latest-bad-news-on-carbon-capture-from-coal-power-plants-higher-costs-51440.

41. Alice Favero, Robert Mendelsohn, et al., "Using Forests for Climate Mitigation: Sequester Carbon or Produce Woody Biomass?" *Climate Economics* 144, no. 2 (September 2017): 195–206.

42. The sink/source terminology is common in IPCC literature.

43. Gregg Marland, *The Prospect of Solving the CO_2 Problem through Global Reforestation* (Washington, DC: US Department of Energy, 1988).

44. See "Impacts World 2017: Counting the True Costs of Climate Change" (conference; Potsdam Institute for Climate Impact Research, Potsdam, Germany, October 11–13, 2017).

45. See Archer, *Carbon Cycle*.

46. *Wikipedia*, s.v. "Albedo," last edited December 1, 2018, https://en.wikipedia.org/wiki/Albedo.

47. Ibid.

48. A climatologist friend once did a quick calculation about clearing the Amazon rainforest and replacing it with grass, as was part of a plot in a draconian environmental disaster book of some decades back. According to the calculations of my climatologist friend, however, this action would result in a net global cooling as the albedo cooling effect would dominate the increased carbon dioxide warming effect by a substantial amount.

49. Jessica Leber, "Scientists Warily Look to Geoengineering to Stave Off Polar Catastrophe," *Oceans Deeply*, December 13, 2017, https://www.newsdeeply.com/oceans/articles/2017/12/13/scientists-warily-look-to-geoengineering-to-stave-off-polar-catastrophe.

50. *Carbon Brief* and Duncan Clark, "How Long Do Greenhouse Gases Stay in the Air?" *Guardian*, January 16, 2012, https://www.theguardian.com/environment/2012/jan/16/greenhouse-gases-remain-air.

51. Jan C. Minx et al., "Negative Emissions—Part 1: Research Landscape and Synthesis," *Environmental Research Letters* 13, no. 6 (May 22, 2018).

52. Chelsea Harvey, "This Is How We Might Pull CO2 Out of the Atmosphere," Governors' Solar Wind & Energy Coalition, Monday, June 5, 2018, http://governorswindenergycoalition.org/this-is-how-we-might-pull-co2-out-of-the-atmosphere/.

53. No-till agriculture involves not tilling the land, which reduces weeds but frees carbon, instead relying on pesticides to control weeds. Thus, the lower disturbance of the soil reduces carbon losses. On many sites this approach reduces costs as well.

54. Archer, *Global Carbon Cycle*.

55. R. A. Berner and A. Lasaga, "Modeling the Geochemical Carbon Cycle," *Scientific American* 260 (1989): 74–81.

56. Archer, *Global Carbon Cycle*.

57. See Favero, Mendlesohn, et al., "Using Forests for Climate Mitigation."

58. Ibid.

59. Harvey, "This Is How We Might Pull CO2."

60. Ibid.

61. Sabine Fuss et al., "Betting on Negative Emissions," *Nature Climate Change* 4 (2014): 850–53.

62. Scott Barrett, *Environment and Statecraft: The Strategy of Environmental Treaty-Making* (Oxford: Oxford University Press, 2006).

63. Scott Barrett, "The Incredible Economics of Geoengineering," *Environmental and Resource Economics* 39, no. 1 (2008): 45–54.

64. Barrett, "Solar Engineering's Brave New World."

65. Klein, *This Changes Everything*.

66. Klein, "Geoengineering."

67. Victor et al., "Geoengineering Option."

68. Ibid.

CHAPTER 6. POLITICAL CHALLENGES

1. L. Kairiukstis, S. Nilisson, and A. Straszak, eds., *Forest Decline and Reproduction: Regional and Global Consequences* (working paper, International Institute for Applied Systems Analysis, Laxenburg, Austria, September 1987).

2. Clean Air Act of 1963, 42 U.S.C. § 7401 (1990).

3. Roger Sedjo, "Pollution Related Forest Decline in the US and Possible Implications for Future Harvests," in *Forest Decline and Reproduction*, eds., Kairiukstis, Nilisson, and Straszak,; Clean Air Act.

4. Kairiukstis, Nilisson, and Straszak, eds., *Forest Decline and Reproduction*.

5. Ozone occurs both at high levels in the stratosphere, roughly ten to thirty miles, and at lower levels, up to about ten miles above the earth's surface. The ozone at lower levels is a destructive pollution, damaging human and animal health. At high levels it forms a shield against ultraviolet radiation, which can cause cancer and disrupt photosynthesis.

6. "About Montreal Protocol," United Nations Environment Programme, http://web.unep.org/ozonaction/who-we-are/about-montreal-protocol.

7. See the Institute for Governance & Sustainable Development, http://www.igsd.org/.

8. Chris Buckley and Henry Fountain, "In a High-Stakes Environmental Whodunit, Many Clues Point to China," *New York Times*, June 24, 2018, https://www.nytimes.com/2018/06/24/world/asia/china-ozone-cfc.html.

9. Scott Barrett, *Negotiating the Next Climate Change Treaty* (London: Policy Exchange, 2010). In this piece Barrett argues for a climate agreement with a structure much closer to that of the Ozone Hole agreement than those that have proposed (p. 25).

10. The characterization of increasing integration is not entirely correct. A move away from internationalism back toward state nationalism can be found in the UK's withdrawal from the EU and growing nationalism in other countries like Turkey.

11. Manipulating a system for a desired outcome.

12. "What Is the Kyoto Protocol?" United Nations Climate Change, https://unfccc.int/process-and-meetings/the-kyoto-protocol/what-is-the-kyoto-protocol/what-is-the-kyoto-protocol.

13. Scott Barrett argued in a seminar at RFF in that era that only one or two European countries would have trouble meeting their Kyoto targets due to preplanned projects.

14. Rules in the Kyoto Protocol called for the projects to be "additive." However, there rules were readily negotiated away.

15. *Monthly Energy Review* (Washington, DC: US Energy Information Administration, November 20, 2018).

16. Hydraulic fracturing for the production of natural gas and oil.

17. Barrett, *Negotiating the Next Climate Change Treaty*.

18. While such an activity might run afoul of its neighbors wishes or possibility violate a UN rule, cooperative action is probably not technically required.

19. Scott Barrett, *Why Cooperate? The Incentive to Supply Global Public Goods* (Oxford; New York: Oxford University Press, 2007).

20. Norman J. Rosenberg, William E. Easterling III, Pierre R. Crosson, and Joel Darmstadter, eds., *Greenhouse Warming: Abatement and Adaptation* (Washington, DC: Resources for the Future, 1989).

21. Figure 2.1.

22. T. M. L. Wigley, et al., "Uncertainties in Climate Stabilization," *Climatic Change* 97, no. 1-2 (2009).

23. Marshall Burke, "Large Potential Reduction in Economic Damages under UN Mitigation Targets," *Nature* 557 (2018): 549–53.

CHAPTER 7. PLAN B AS INSURANCE

1. "California Advances Climate Insurance Effort," Climatewire, June 21, 2018.

2. "The Social Cost of Carbon: Estimating the Benefits of Reducing Greenhouse Gas Emissions," US Environmental Protection Agency, last updated January 9, 2017, https://19january2017snapshot.epa.gov/climatechange/social-cost-carbon_.html.

3. Ibid.

4. T. A. Boden et al., "Global, Regional, and National Fossil-Fuel CO_2 Emissions" (Oak Ridge, TN: Carbon Dioxide Information Analysis Center, Oak Ridge National Laboratory, US Department of Energy, 2017). Estimates of global emissions in 2015 were about 35 billion tons of carbon dioxide: Jos G. J. Olivier et al., *Trends in Global CO_2 Emissions: 2015 Report* (The Hague: PBL Netherlands Environmental Assessment Agency/Joint Research Centre, 2015).

5. Glen P. Peters et al., "The Challenge to Keep Global Warming Below 2 °C," *Nature Climate Change* 3 (2013): 4–6.

6. 1 billion = 1,000 million.

7. F. Moore, "Valuing Climate Damages at the Country Level," *Nature Climate Change* 8 (2018): 856–57.

8. Technically, these estimates are the discounted present value of current and future damages associated with a ton of CO_2 or its equivalent.

9. See National Academies of Science, Engineering, and Medicine, *Valuing Climate Damages: Updating Estimation of the Social Cost of Carbon Dioxide* (Washington, DC: National Academies Press, 2017). The process looks at the damages through time and uses a discount rate to estimate the present value of those damages. There are serious disagreements as to what is the appropriate discount rate for evaluating long-term climate change and associated damages.

10. Simon Evans et al., "Q&A: The Social Cost of Carbon," Carbon Brief, February 14, 2017, https://www.carbonbrief.org/qa-social-cost-carbon.

11. See Gary W. Yohe and Richard S. J. Tol, "The Stern Review: A Deconstruction," (working paper, Research Unit Sustainability and Global Change, Hamburg University, Hamburg, 2007).

12. This value estimate depends on the discount rate chosen.

13. David R. Henderson, "A Nobel Economics Prize for the Long-Run," *Wall Street Journal*, October 8, 2018, https://www.wsj.com/articles/a-nobel-economics-prize-for-the-long-run-1539039555.

14. *Wikipedia*, s.v. "Insurance," last edited December 6, 2018, https://en.wikipedia.org/wiki/Insurance.

15. Fracking (hydraulic fracturing) is the process of injecting liquid at high pressure into subterranean rocks, boreholes, etc. to force open fissures and extract oil and gas. Although this technique is often criticized for environmental damages—exposing potentially dangerous chemicals to the surrounding environment, releasing methane and other pollutants into the air, contamination and heavy use of water resources, increased likelihood of oil spills and earthquakes—it is increasingly practiced in the US and is responsible for the large increase in US gas and oil production over the past decade or so. "What Are the Effects of Fracking on the Environment?" Investopedia, July 16, 2018, https://www.investopedia.com/ask/answers/011915/what-are-effects-fracking-environment.asp

16. G. Bala, "Problems with Geoengineering Schemes to Combat Climate Change," *Current Science* 96, no. 1 (2009): 41–48.

17. See Chapter 5.

18. D. B. Keller et al., "Potential Climate Engineering Effectiveness and Side Effects during a High Carbon Dioxide-Emission Scenario," *Natural Communications* 5, no. 3304 (2014); also N. E. Vaughan and T. M. Lenton, "A Review of Climate Geoengineering Proposals," *Climate Change* 109 (2011): 745–90.

CHAPTER 8. WHERE FROM HERE?
NEXT STEPS AS WELL AS
SUMMARY AND CONCLUSIONS

1. Paul Krugman, "Earth, Wind, and Liars," *New York Times*, April 16, 2018, https://www.nytimes.com/2018/04/16/opinion/trump-energy-environment.html.

2. Daniel Raimi and Alan Krupnick, "Decarbonization: It Ain't That Easy," *Resources for the Future* (blog), April 20, 2018, http://www.rff.org/blog/2018/decarbonization-it-ain-t-easy.

3. K. Jan Oosthoek, "Little Ice Age," Environmental History Resources, June 5, 2015, https://www.eh-resources.org/little-ice-age/.

4. M. Staubwasser et al., "Impact of Climate Change on the Transition of Neanderthals to Modern Humans in Europe," *Proceedings of the National Academy of Science* 115, no. 37 (August 27, 2018): 9116–21, discuss cold periods thought to be associated with the disappearance of Neanderthals 40,000 years ago.

5. Lorraine E. Lisiecki and Maureen E. Raymo, "A Pliocene-Pleistocene Stack of 57 Globally Distributed Benthic δ18O Records," *Paleoceanography* 20, no. 1 (2005); R. Petit et al., "Climate and Atmosphere History of the Past 420,000 Years from the Vostok Ice Core, Antarctica," *Nature* 399 (June 3, 1999). Debates continue over the question of whether GHGs precede warming or warming predate GHG buildups. Certainly there is evidence for warming preceding GHG emissions, as warming is known to free carbon dioxide and methane for permafrost, and there is evidence that certain GHGs have a radiative forcing effect.

6. See Andrew King et al., "What Is a Pre-Industrial Climate and Why Does It Matter?" The Conversation, June 8, 2017, https://theconversation.com/what-is-a-pre-industrial-climate-and-why-does-it-matter-78601.

7. For example, evidence of this skepticism is revealed in the recent outs of the IPCC. See Chris Mooney et al., "The Recent IPCC Report Is Particularly Bleak," *Washington Post*," October 10, 2017.

8. Sea level can be monitored by satellite.

9. See Mooney, et al. "Recent IPCC Report."

10. Long Cao et al., "Simultaneous Stabilization of Global Temperature and Precipitation through Cocktail Geoengineering," *Geophysical Research Letters* 44, no. 14 (July 24, 2017): 7429–37.

11. Brian Wang, "Point of No Return for Climate Action Is Point When Geoengineering Will Start," Next Big Future, September 2, 2018, https://www.nextbigfuture.com/2018/09/point-of-no-return-for-climate-action-is-point-when-geoengineering-will-start.html.

12. *Geoengineering the Climate: Science, Governance and Uncertainty* (London: Royal Society, September 1, 2009).

13. Scott Barrett, *Why Cooperate? The Incentive to Supply Global Public Goods* (Oxford; New York: Oxford University Press, 2007), provides a rationale for when cooperation is desired and when it may not be needed.

14. *US Crude Oil and Natural Gas Proved Reserves, Year-End 2017* (Washington, DC: US Energy Information Administration, November 29, 2018).

15. Again, note that although fracking is controversial and banned in some states

(for its various potential damages to the surrounding environment including air and water pollution) it is being widely practiced and is responsible for the US achieving unprecedented levels of gas and oil production.

16. Timothy Puko, "Power Plant Rules Set Up Legal Fight," *Wall Street Journal*, August 21, 2018, P A3. Most of the reductions in US carbon dioxide emissions can be attributed to the substitution of natural gas for coal in the power industry.

17. Although coal is being displaced in the US and much of the developed world, this is much less true in the developing world.

18. Christopher Solomon, "Alaska Wants to Build a Second 800-Mile Pipeline," *Outside*, May 29, 2018.

19. Adam Vaughan, "Fracking: The Reality, the Risks and What the Future Holds," *Guardian*, February 26, 2018, https://www.theguardian.com/news/2018/feb/26/fracking-the-reality-the-risks-and-what-the-future-holds.

20. *Intergovernmental Panel on Climate Change Special Report: Global Warming of 1.5 °C* (Geneva, Switzerland: Intergovernmental Panel on Climate Change, October 2018).

INDEX